OXFORD GEOGRAPHICAL AND ENVIRONMENTAL
STUDIES

Editors: Gordon Clark, Andrew Goudie, and Ceri Peach

ENERGY STRUCTURES AND ENVIRONMENTAL FUTURES

Energy Structures and Environmental Futures

Torleif Haugland, Helge Ole Bergesen,
and Kjell Roland

OXFORD UNIVERSITY PRESS
1998

Oxford University Press, Great Clarendon Street, Oxford OX2 6DP

Oxford New York

Athens Auckland Bangkok Bogota Bombay
Buenos Aires Calcutta Cape Town Dar es Salaam
Delhi Florence Hong Kong Istanbul Karachi
Kuala Lumpur Madras Madrid Melbourne
Mexico City Nairobi Paris Singapore
Taipei Tokyo Toronto Warsaw

and associated companies in
Berlin Ibadan

Oxford is a trade mark of Oxford University Press

Published in the United States
by Oxford University Press Inc., New York

British Library Cataloguing in Publication Data

Data available

Library of Congress Cataloging in Publication Data
Haugland, Torleif.
Energy structures and environmental futures in Europe/
Torleif Haugland, Helga Ole Bergesen, and Kjell Roland.
p. cm.—(Oxford geographical and environmental studies)
Includes bibilographical references (p.).
1. Energy development—Europe. 2. Energy policy–Europe.
3. Environmental policy—Europe. I. Bergesen, Helga Ole.
II. Roland, Kjell. III. Title. IV. Series.
TJ163.25.E85H39   1998   333.79'094—dc21   97—39486

ISBN 0–19–823360–4

1 3 5 7 9 10 8 6 4 2

Typeset by J&L Composition Ltd, Filey, North Yorkshire
Printed in Great Britain by
Bookcraft Ltd, Midsomer Norton, Somerset

EDITORS' PREFACE

Geography and environmental studies are two closely related and burgeoning fields of academic enquiry. Both have grown rapidly over the past two decades. At once catholic in its approach and yet strongly committed to a comprehensive understanding of the world, geography has focused upon the interaction between global and local phenomena. Environmental studies, on the other hand, have shared with the discipline of geography an engagement with different disciplines, addressing wide-ranging environmental issues in the scientific community and the policy community of great significance. Ranging from the analysis of climate change and physical processes to the cultural dislocations of post-modernism and human geography, these two fields of enquiry have been in the forefront of attempts to comprehend transformations taking place in the world, manifesting themselves in a variety of separate but interrelated spatial processes.

The new 'Oxford Geographical and Environmental Studies' series aims to reflect this diversity and engagement. It aims to publish the best and original research studies in the two related fields and in doing so, to demonstrate the significance of geographical and environmental perspectives for understanding the contemporary world. As a consequence, its scope will be international and will range widely in terms of its topics, approaches, and methodologies. Its authors will be welcomed from all corners of the globe. We hope the series will assist in redefining the frontiers of knowledge and build bridges within the fields of geography and environmental studies. We hope also that it will cement links with topics and approaches that have originated outside the strict confines of these disciplines. Resulting studies will contribute to frontiers of research and knowledge as well as representing individually the fruits of particular and diverse specialist expertise in the traditions of scholarly publication.

Gordon Clark
Andrew Goudie
Ceri Peach

ACKNOWLEDGEMENTS

This book is the result of a research project financed by the Norwegian Research Council under the programme Society, Energy, and Environment (SAMMEN). Three institutions have participated in the project: Fridtjof Nansen Institute (FNI), ECON Centre for Economic Analysis (ECON), and Energidata. Although Torleif Haugland (ECON), Helge Ole Bergesen (FNI), and Kjell Roland (ECON) are the authors of the book, several researchers from the three institutions made important contributions. Erik Sørensen (ECON) wrote large parts of Chapters 4, 5, and 8. Erich Unterwurzacher (ECON) contributed to Chapters 2 and 6. Javier Estrada (FNI) contributed to Chapter 3. Geir Falkenberg (FNI) and Jan Arild Snoen (ECON) contributed to Chapter 7 and Agnete Dahl to Chapter 8. Oddbjørn Fredriksen and Arne Ljones (Energidata) have contributed to section 2.2.

During the course of the work several workshops have been held to discuss drafts of the manuscript. Many people have given generously of their time to read and comment on drafts that did not always make a pleasant read. The authors would in particular like to thank Per Ove Eikeland, Javier Estrada, Leiv Lunde, Anne Kristin Sydnes (FNI), Paul Parks and Per Schreiner (ECON), Michael Grubb and Jonathan Stern (Royal Institute of International Affairs), Lutz Mez (Free University of Berlin), Kevin Leydon (the European Commission), Dominique Finon (Institut d'Economie et de Politique de l'Energie, Grenoble), Atle Midtun (Oslo Business School), Trond Kubberud (the Norwegian Ministry of Oil and Energy), and Stein Hansen, chairman of the board for the SAMMEN programme. Invaluable research assistance and technical support have been provided by Susan Høivik, Jan Arild Snoen, Heidi Bråten, Hege Løvås, and Nina Strømme.

T.H., H.O.B., K.R.

CONTENTS

LIST OF FIGURES

LIST OF TABLES

ABBREVIATIONS

AIJ	Activities Implemented Jointly
BCM	billion cubic metres (10^9 m^2)
boe	barrel of oil equivalent
CAFE	corporate average fuel economy
CCGT	combined-cycle gas turbines
CdF	Charbonnages de France
CEGB	Central Electricity Generating Board
CIF	cost insurance freight
CMEA	Council for Mutual Economic Assistance
DEM	Deutsches Mark
DSM	demand-side management
EdF	Electricité de France
ENEL	Ente Nazionale per l'Energia Elettrica
ENI	Ente Nazionale Idrocarburi
EU	European Union
FCCC	Framework Convention on Climate Change
GW	gigawatt (10^9 watt)
GWh	gigawatt hour
IEA	International Energy Agency
IEM	Internal Energy Market
IPCC	Intergovernmental Panel on Climate Change
kWh	kilowatt hour
LNG	liquefied natural gas
mbd	million barrels a day
MMbtu	Million British thermal unit
MMC	Monopoly and Mergers Commission
mtce	million tonnes of coal equivalent
mtoe	million tons of oil equivalents
MW	megawatt
PSE	Producer Subsidy Equivalent
R&D	Research and Development

REC	Regional Electricity Company
SEA	Single European Act
SEP	Samenwerkende Electriciteits-Produktiebedrijven
tce	tonne of coal equivalent
TPA	third-party access
TPES	total primary energy supply
TWh	terawatt hour
UN ECE	United Nation Economic Commission for Europe

1
Introduction

If anything is inchoate it is energy futures
(Schwarz and Thompson, 1990: 97)

Without modern energy systems, society as we know it today would cease to exist. Providing adequate energy supplies at reasonable prices has been an integral part of modernization and nation-building throughout the western world. Even if modern engineering has been essential, energy has invariably been considered too important to be left to engineers alone. Politicians, industrialists, trade unions have all made their mark on energy development, which has therefore been shaped by changing societal concerns and political priorities over time, and not least by distinct national differences. Energy is not only a basis for modern society, but also a product of it. This book is a study of the close and ever-changing relationship between the energy sector and the society that surrounds it.

At the end of the twentieth century this relationship faces two fundamental challenges: First, the national confinement of modern energy systems is undermined by technological progress, making long-distance trade increasingly attractive, and by the broad trend towards economic internationalization in general and political integration in Europe in particular. Second, the risk of climate change may lead governments and publics to demand a profound restructuring of the entire energy sector.

Our purpose is to analyse how these two fundamental challenges, and the connection between them, can affect future energy developments in Europe. The analysis must be rooted in a firm understanding of the past. The first part of the book is therefore devoted to a systematic description and analysis of the energy sector in Europe as it has developed over the past twenty-five years, by major subsectors and with examples from the most important countries. In this retrospective section our intention is to develop a straightforward framework of analysis that combines economics and political science, within which to demonstrate in the empirical chapters both the remarkable continuity in energy development in our part of the world and the undercurrents that threaten the status quo in the late-1990s.

Only when understanding of the past is firmly established is it fruitful to proceed to predictions of the future. The literature of the 1990s abounds with scenarios of energy futures, most of them based on presumptions which are either rather arbitrary, being developed from political objectives, or mainly technical. The former can easily lead to wishful thinking; the latter will often neglect current realities.

We have taken a different approach. The scenarios in Part II are rooted in our analyses of energy structures in Part I. Whatever major changes may be in the offing in the early twenty-first century, they will have to work through the structures of the twentieth, and these are not going to yield easily either to economic internationalization or political environmentalism. We want to know, more precisely, how current energy structures will interact with these two broad currents of change. Is it possible to combine the set-up of the 1990s with a quantum leap in economic and political integration in Europe? How will the energy sector be affected by and react to the demands following from such a radical change?

If climate change—probably at the top of a new wave of environmentalism—becomes an overriding political priority, how can this potentially revolutionary force act through energy structures which are either an extension of the present or undergoing a transformation from the national to the European level of government? Again, the future cannot be predicted without a firm grasp on the present and a thorough analysis of the past.

1.1. The Post-War Agenda

Shifting Priorities in Energy Policy

Concern with energy supplies dates back at least to 1865, when Jevons predicted the end of the Industrial Revolution as England would run out of coal. In most societies access to cheap and reliable sources of energy has been and still is a major public concern. But governments also have other interests in energy, as we can see from the evolution of the energy policy agenda since 1945. Since the Second World War, three major issues have shaped the direction of energy development in Western Europe.

Reconstruction, Industrialization, and Economic Growth

The immediate post-war period was characterized by government planning and control of most economic activities. The pivotal role played by the energy sector in reconstruction and industrialization made energy policy a

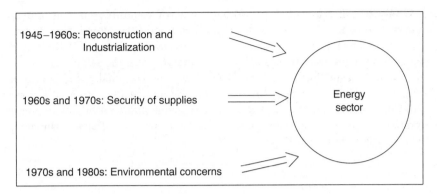

Fɪɢ. 1.1. The post-war energy policy agenda

major priority. In some countries where large national monopolies were created through nationalization of energy enterprises, public ownership became an important vehicle for attaining energy policy objectives. In other countries, a decentralized company structure remained in place, but intimately linked with the public sector and under tight governmental control and supervision.

In the 1950s and 1960s, the predominant issue in energy policy was how to provide sufficient energy to fuel rapid economic growth. The technological and financial challenges in building adequate energy-supply capacity were immense. Major investments were required to develop generating capacity and infrastructure to meet the rapidly growing demand for electricity. This period saw the emergence of large and often vertically integrated energy enterprises able to master modern capital-intensive, large-scale technologies.

Security of Supply

In the late 1960s and in the 1970s, the geopolitical aspects of energy supplies came to the forefront. Growing trade in oil and natural gas called for national control of the supply chain for energy traded internationally. The oil sector became a target of governmental intervention. In many countries, governments took ownership or actively supported the formation of domestic companies to engage in exploration for oil and natural gas, as well as in downstream operations such as imports, refining, and distribution. Initially these steps were taken in order to reduce the dominance and profit of vertically integrated (British and US) multinational oil companies. The oil crises of the 1970s reinforced and established in an entire generation of policy-makers a world-view centred on energy security. Both balance-of-payment and security-of-supply considerations led to a strengthening of governmental control over imports and exports, storage, distribution, and pricing of oil and natural gas.

Furthermore, the oil-price shocks, combined with public R&D (Research and Development) initiatives and political influence over the electricity sector, led to a significant shift in fuel mix, away from oil. This change took place in Western Europe in the late 1970s and throughout the 1980s. In Central Europe it did not begin until the mid-1980s. (Throughout this book we use Central Europe, or the Visegrad countries, as a generic term for Poland, Hungary, Slovakia, and the Czech Republic.)

Environmental Concerns

Environmental issues have gradually grown in influence on energy policy. The rise in popular pressure to preserve the natural environment is a strong indication of this trend. At an early stage, the construction of hydropower dams met with resistance—however, often without leading to major alterations in project implementation. Another issue was concern over the effects on marine life of oil spills both in open water and in coastal areas. Controversies over the siting of thermal power plants and transmission lines also became widespread, creating major complications for investment programmes in the power sector. Protests against nuclear power became organized in the early 1970s. Health and safety risks related to nuclear power became a top-priority concern after the Three Mile Island accident in 1979, and new ventures came to a standstill in most countries after the Chernobyl disaster in 1986.

In the 1970s, energy-related air pollution started to impact on policy. Initially the focus was on local air quality. In the 1980s, transboundary problems related to emissions of acid rain precursors such as SO_2 and NO_x represented new challenges for power plants, oil refineries, and other industries. These issues were in most countries dealt with at the national level, usually through a case-by-case approach, mandating technological measures or product and emission standards.

From this brief historical sketch we can see that policies and priorities directed towards the energy sector have changed substantially in recent decades. In some cases, political shifts have occurred abruptly, with little warning and with profound effects on energy markets. Despite these alterations in policies and priorities, the general institutional patterns and the overall regulatory system of European energy sectors have changed only very slowly, remaining largely unaltered for extended periods of time. Public ownership and tight regulation have continued to be the principal means of conducting energy policy. Even the introduction of a new energy product—natural gas—

could be absorbed by the markets without major shake-ups in existing energy industries.

Until the 1990s, the structure of the energy sector and government/industry relationship in Western Europe outside the UK had not been seriously called into question. One important reason for this was that the energy sector, within the framework of existing institutional patterns and regulatory systems, had adapted to the main political priorities of society. When new challenges arose, the energy sector was able to address them and respond. In general, the end result was satisfactory to governments and electorates. The resilience of energy sector institutions, together with the effective interplay between operating entities and policy bodies in nationally protected and rapidly expanding markets, can explain the flexibility of the response to new challenges. Another reason is the sequential manner in which the main political priorities surfaced. As a rule, the sector was exposed to one coherent set of policy priorities at a time. Thus, the direction in which to move and the thrust of the measures to be undertaken were, if not obvious, at least clearly discernible to most of the actors involved.

1.2. The New Agenda

Pressures for Change

Since the mid-1980s, the strategic challenges facing the energy sector have gradually become more complex and onerous. Mention has already been made of the major unresolved environmental problems related to energy supply and consumption. The threat of global warming has now moved to the top of the agenda in both international and national politics. In addition, several other political and economic high-priority issues with potentially large impacts for the energy sector have surfaced. All these new challenges call into question today's national confinement of energy markets and energy/environmental policies. Despite a free flow of technology, extensive trade in energy and shifting political priorities in the post-war period, the energy sector has remained entirely in the hands of nation-states. In many respects, differences in industrial structure, fuel mix, and technology are more easily explained by reference to the country of location than by the particular energy industry in question. This strong national bent in policies and industries has come under severe strain.

We have identified five major challenges that could easily be transformed into forces capable of undermining the current structure of the energy sector.

These constitute the agenda that will dominate energy development in Europe
in the years to come:

- Increasing environmental concerns, in particular climate change
- Economic integration in Western Europe
- The transition from planned to market economies in the East
- New technological options
- Deregulation and competition

This new agenda does not imply that the old one has disappeared. Reliable
and predictable supplies will remain a concern for all governments. An
energy sector that neglects this objective will not be acceptable in the long
term, regardless of ownership, structure, or whatever policy issue may be
high on the agenda. Security of supply is a prerequisite for stability and
acceptability.

Due to the size of the sector in terms of employment, turnover, and capital,
strong social interests will maintain a place on the policy agenda. However,

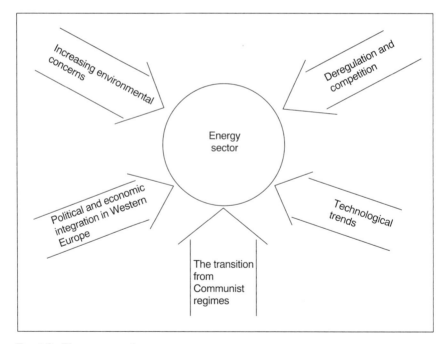

F<small>IG</small>. 1.2. The new agenda

their importance will change over time, depending on the structure of the sector and on general ideological trends. In the past, employment and regional policy have played a crucial role. In the future, important linkages to agricultural policy might develop through production of biomass energy; or efficiency and costs could become important if consumer interests move to the forefront.

The Environment—the Second Wave

Attention to environmental problems developed slowly in the late 1960s as the public and politicians gradually became aware of the ecological costs of increasing affluence. In this early period, interest focused primarily on tangible and geographically confined issues: pollution of rivers, lakes, and oceans; contamination of water and soil; noise and air pollution in cities. Problems were defined largely in terms of point-source emissions—particularly industrial sources—which at that time could most readily be tackled by end-of-pipe solutions imposed upon polluters through national regulation.

International concern, largely confined to the West, culminated with the first UN conference on the environment held in Stockholm in 1972. The public concern stimulated by this event led to the establishment of several non-governmental organizations (NGOs) of lasting importance, like Friends of the Earth and Greenpeace. This first wave of environmentalism saw regulatory agencies, often with the status of government ministries, set up all over the OECD area. Shortly thereafter national regulators embarked on the long road from policy formulation, through hearings and legislation, towards implementation. The process was often slow, and the pace varied from country to country, but the thrust was clear: the squeeze on polluting industries became harder and harder. Emissions of major pollutants began to fall noticeably from the late 1970s onwards (cf. Chapter 7).

In the international arena, however, progress was much slower. For a long time, transnational issues like acid rain suffered from neglect by most governments and from lack of institutional focus. Nevertheless, cooperative schemes gradually developed—joint research programmes, monitoring devices—and a common scientific understanding of ecological problems of a transboundary nature gradually emerged.

Although public concern with the environment declined in the late 1970s and into the early 1980s, the institutional set-up remained in place: national agencies and programmes with fairly wide coverage as far as pollution was concerned, rudimentary frameworks for international cooperation, and a certain level of grass-roots organization. The second wave of environmental concern got under way in the mid-1980s, from the platform created by its predecessor. This goes a long way to explaining the speed and strength of the second strand of environmentalism: Within a few years, environmental issues had risen to the top of the political agenda, both at the grass-roots level and in governmental circles. This led to renewed demands for action from NGOs and from newly founded green political parties in some countries. Politicians in Europe and North America soon latched on to the new trend, joining the green 'beauty contest', nationally as well as internationally. Ministries and agencies responsible for the environment regained momentum, soon finding themselves able to win support for radical emission reductions and a renewed squeeze on the remaining large polluters.

Internationally, the change from the 1970s and early 1980s was even more pronounced. Negotiations long bogged down in technical disputes suddenly gained a new lease on life. The laborious process of regulating emissions in Europe started rolling with new speed, leading to rapid ratification of two protocols on SO_2 and NO_X under the auspices of UN ECE. At the same time the governments of the European Community negotiated and signed an agreement regarding emissions of SO_2 and NO_X from power plants and large industrial boilers, the Large Combustion Plant Directive. Negotiations on ozone depletion, long stalled over a serious US–EC dispute, led in record time to the Montreal Protocol of 1987, with subsequent amendments.

The major issue on the renewed environmental agenda, essential in this context, is climate change. Again there was a slow, almost invisible, intergovernmental process that gained surprising momentum by the late 1980s. This first led to the establishment of the Intergovernmental Panel on Climate Change (IPCC) in 1988. Later the Framework Convention on Climate Change (FCCC) was signed in Rio de Janeiro in 1992 and entered into force in 1994. In 1995, IPCC produced its second scientific assessment, confining the area of uncertainty considerably compared with the first, thereby adding to the pressures for international political solutions (IPCC 1990 and 1996). Compared with similar processes in the 1970s, the speed and force of intergovernmental environmentalism in the early 1990s have been astounding.

To the energy sector, climate change is a fundamental challenge in two ways. First, a policy that is reasonably successful in curbing emission of greenhouse gases will have to be coordinated and implemented at the international level. Today's national institutions have proven inadequate to deal with the issue in an effective and efficient manner. Secondly, if effective measures to curb emission are implemented, this will radically change the existing structure of energy industries and markets.

Political and Economic Integration in Western Europe

In 1987 the EC launched the Single European Act (SEA) as a move towards greater competitiveness and the creation of a single, strong European market to revitalize the economies of Western Europe. Coming after many years of stalemate, the SEA was triggered by political concern over Europe's poor economic performance. The general view was that this was rooted in lack of economic flexibility, slow technological progress, and small and fragmented markets—the aggregate effect often referred to as 'eurosclerosis'.

An important component of the SEA was to apply the competition rules of the Treaty of Rome to previously excluded sectors such as energy, transport, telecommunications, and public procurement. Hence energy reform was elevated to an important political issue, with the objective of creating an internal energy market (IEM). The basic idea was to liberalize trade and thus create pressure on the separate and often diverging national energy policies within the Community. In turn, increasing energy interdependence was also expected to stimulate policy-making at the EC (later EU) level. Market integration was to pave the way for political integration and run parallel to it. As the internal energy market develops, the role of member states in securing energy supplies and implementing environmental policy will be constrained. Instead, the European Union will have to adopt policies which can cover the entire internal market, i.e. a common EU energy and environmental policy.

To energy industries, an internal market or the status quo, in terms of nationally confined markets, represent two very different states of the world. An open and competitive market introduced over a short period of time would create havoc for many companies now engaged in protected and monopolized national markets. For other companies, major new business opportunities would arise.

The Transition from Communist Regimes

The political map of Europe has changed, and the new landscape will have important implications for energy and environment. The collapse of Communism instantly altered the conditions for political and economic cooperation in Europe. The system of central planning and economic cooperation within the Communist bloc had created economic structures and trade flows in Central Europe that represented a massive waste of capital, and natural and human resources. The transition towards democracy and a modern market economy means profound changes in old political, social, and economic structures.

Energy was an important component of the old economic order. Underpriced energy flowed from the Soviet Union to its East European satellites, while manufactured products, some of them energy-intensive, were shipped east-wards in exchange. At the national level, investments in coal production were pursued far beyond the economic optimum in order to limit reliance on energy imports. Inefficient investment strategies were followed in other energy sectors as well, particularly in electricity and heat supplies. The performance of energy-supply industries was further hampered by lack of a clear and trans-parent regulatory system and by poorly developed economic incentives to conserve resources. Energy-sector enterprises were organized as oversized State Trusts, with no distinction made between commercial/technical opera-tions and regulatory/political dispositions. The result was poor operational management, neglect of maintenance and repair, and slow technological pro-gress. At the end-use level, major energy wastage prevailed due to low prices, soft budget constraints for industry, and lack of technical devices and incen-tives to optimize consumption. Energy intensity (energy consumption per unit of GDP) in Central Europe is still two to three times higher than in Western Europe. The present share of coal and lignite in end-use consumption in the former Eastern bloc resembles the structure in Western Europe of the 1950s.

On the ruins of this perverted energy system, new market structures and economic relations are now in the making. In the Central European countries there is a general awareness that the energy sector will have to change radically if the requirements of a market economy are to be met. The ultimate goal may be to privatize a major part of the energy sector, but it is not yet clear what the full range of the reform process will be. The petroleum sector has been deregulated, and important steps have been taken to give enterprises greater operational and commercial autonomy from state intervention. More-over, new legislation and regulatory systems for the grid-based energy sectors (natural gas, electricity, and heat) have been established in most Central European countries, but with unresolved questions related to ownership and company structure for the commercial part of energy supplies.

The basic issues for the years ahead concern the role of politics versus markets in price setting, financial constraints, and, not least, co-operation with the West. Will Central Europe retain its isolated national energy markets or join an expanding energy trading area in Western Europe?

New Technological Options

For decades, technological progress has contributed to major productivity gains in energy supplies and improved efficiency in energy use. Any study on future energy developments will need to bear in mind the crucial role of technology. Technological trends are important not only for technical efficiency in supply and consumption, but also for industrial structures and for the design of regulatory regimes and policy.

The impact of technology on energy markets has profoundly affected the costs and quality of supplying energy services. One kilowatt-hour of electricity can today be delivered to individual households at a much lower price, with greater regularity, with less variance in frequency, and—importantly—often with less harm to the environment than only two or three decades ago. Technological progress has enhanced the use of electricity in end-uses previously served by coal, wood, or petroleum, increasing comfort and reliability. New technologies have brought to the market new forms of energy, such as wind power and solar energy. To judge by current R&D efforts, future energy options may include hydrogen, fuel cells, improved nuclear-power designs, and, not least, improvements in conventional technologies that can drastically alter their operational characteristics.

As technology progressed in the post-war period, there was an increase in the optimal size of units in energy production, particularly in electricity generation. The organizational and structural counterpart to economies of scale was the growth of large integrated companies which were granted *de facto* or *de jure* monopoly in markets for electricity and gas. This in turn was paired with tight public control. The capital intensity of production and monopolistic structures in the vertical chains acted to deter new entrants.

Recent technological developments have begun to challenge this pattern. Gas-fired technologies in power production (for example, the combined-cycle gas turbines) capture economies of scale in comparatively small units. Both capital costs and environmental attractiveness (including reduced siting difficulties) make them competitive with, for example, larger coal-fired plants. Co-generation of heat and power in modular small-scale energy systems in the hands of consumers may in the near future gain in economic attractiveness. Solar and wind power owned and operated by consumers or small-scale independent producers can also come to challenge the power of electric utilities.

Modern information technology reduces economies of scope in grid-operated energy systems. Utilities traditionally sell a bundle of services: electricity, transmission, distribution, measuring, and billing. Formerly, this could be done more cheaply by one company. This is no longer the case: computer technology has made possible unbundling of services by moving large volumes of information around at low cost. In technical terms, even individual households can now shop around the market for the cheapest electricity supplies, in a way inconceivable only a few years ago.

A trend towards smaller units in energy supply will undermine the dominant position of large and often vertically integrated energy enterprises. Increases in co-generation with surplus energy available to the grid will in many countries call for regulatory changes that will affect the interface between grid-owners and independent producers. If the trend towards 'clean and cheap' small-scale technologies continues at today's rapid pace, this would represent a major force for radical changes to the structure of the energy sector.

Further development and improvement of existing technologies is also likely to change their relative attractiveness in terms of both costs and environmental impacts. Clean-coal technology is within reach, and widespread use is likely over the next decade without adding to the cost of electricity generation. The costs of solar power and other renewables are coming down rapidly, which could make such technologies commercially attractive in many end-uses early in the next century. The most far-reaching perspective on the horizon today is the possibility that the menu of technologies available in the future could undermine the firmly held tenet in energy industries that big is beautiful and cheap.

Deregulation, Privatization, and Competition

Deregulation has become the common term to describe the trend in energy policy which started during the Carter Administration in the USA, gained momentum under Reagan, as well as under Thatcher in the UK, and eventually from the late 1980s spread throughout the world. The main economic objective of this trend is to establish a regulatory framework which relies more heavily on market forces and incentives to increase efficiency in energy markets. The term 'deregulation' is somewhat misleading, since very often the volume of regulations actually increases. A more appropriate label would probably be 'reregulation'. The new regulatory philosophy has been to instil competition through corporativization and privatization, the break-up of public monopolies, the separation of competitive and monopolistic activities, and the removal of barriers to entry. Deregulation has involved many sectors of the economy, but energy distribution, transportation, and communications are among the most affected.

The driving forces behind regulatory reforms vary from country to country. In the USA, this trend emerged in the late 1970s as a response to an energy sector completely devastated by bureaucracy, with layer upon layer of extremely complicated and often meaningless regulations. Regulatory failure was so oppressive as to seriously affect daily operations of the energy sector. By contrast, in the UK reforms were motivated mainly by a general ideological belief in free markets and reduced government intervention in business operations. Another motive was the fight against vested interests capturing the sector, in particular public-sector trade unions.

The question of the full effects and the merits of deregulation is a critical subject of intense debate, especially in relation to the reforms undertaken in Norway and United Kingdom, and with EU attempts to establish an internal energy market through supranational regulations. Stimulated by a relaxed attitude to security of supply, perceptions of feasible ways of organizing the energy sector have changed substantially over the last few years. If the ideas tested out in the UK, Norway, and some other countries are judged to deliver what is promised, European energy sectors may be in for some far-reaching changes in the next decade.

Again, an important point is the relation between national markets and regional aspects. The case of electricity reform in Norway, and more recently in Sweden, shows that deregulation is difficult to implement in one country when extensive trading arrangements are maintained with still-monopolized markets. The interface between widely different systems creates tensions.

1.3. Framework of Analysis: A Multidisciplinary Approach

Taken separately, each of the five forces that constitute the new agenda could affect the energy sector profoundly. In analysing the impact of only one force, we need to address a range of technological, economic, institutional, and political factors. When all of these forces are brought to bear on the sector at the same time, the analytical challenge becomes even more complex. Many observers and analysts respond to this complexity by seeking to reduce the chaotic picture to one overriding dimension, such as: energy development is primarily driven by market forces/by new technological options/by security of supply policy. Our point of departure is that any such attempt at identifying *the* major factor is futile and bound to be misleading. Such an approach might have been acceptable in the past when the sector was exposed to a policy agenda that changed in a sequential manner, with one dominant issue topping the agenda at a time. Today, such approaches are inappropriate. There is no alternative but to face the inherent complexity head-on. We have therefore developed a framework of analysis deliberately designed to take into account the interdisciplinary nature of our subject-matter, the complex linkages between politics and markets in energy development in Europe.

We start from the assumption that a fruitful analytical approach must go beyond the confines of individual scientific disciplines. The traditional *economic approach* to energy-sector analysis begins with the assumption that government institutions and policies are exogenous to the market, that explicit targets are specified, and that certain policy instruments are available to policy-makers. Economic analysis then explains how the market will evolve over time, given a certain macro-economic development and a specific set of policies adopted by governments. Demand follows from the level of economic activity and the prices charged for different forms of energy. Composition of supply is determined by relative costs of providing energy to different end-users, including the cost of conversion, transportation, and government taxes. In the medium term, inefficiencies, politics, and structural deficiencies (e.g. market power) can create deviation from what is economically optimal, but in the long term, a rational economic system is bound to emerge.

The *political science approach* does not assume any inherent economic rationality. On the contrary, political scientists will often focus their analyses on attempts by vested interests to intervene in markets to promote their own position. Energy markets are studied as a political arena where diverse actors, including political parties, trade unions, corporations, and public bureaucracies, compete for political power and economic advantage. This can easily lead the analyst to neglect the force of underlying market dynamics, the nuts and bolts of economic analysis. Finally, the *engineering approach* is to explore the technically optimal way of designing and operating a system. Less attention is, however, paid to economic and political feasibility.

Our approach differs from all of these in the sense that we recognize that the evolution of the energy sector is shaped both by market forces and the interplay among corporate and political interests both inside and outside the sector. The development is influenced by political decisions and ideological trends in society at large, by vested interests within the sector, as well as the management responsible for the day-to-day operation of the whole range of commercial and semi-commercial entities that together comprise the sector. No a priori assumptions are made with respect to whether energy markets are driven primarily by economic, political, or technological forces. We are convinced that the key to understanding both the past and the future lies in rejecting the narrow unidimensional approaches of traditional analysts.

In our framework of our analysis we distinguish between three broad categories of factors that affect energy markets (see Fig 1.3): the economy at large, energy and environmental policies, and the structure of energy industries. We are particularly interested in the latter two and the interaction between them that is explained in terms of a social contract.

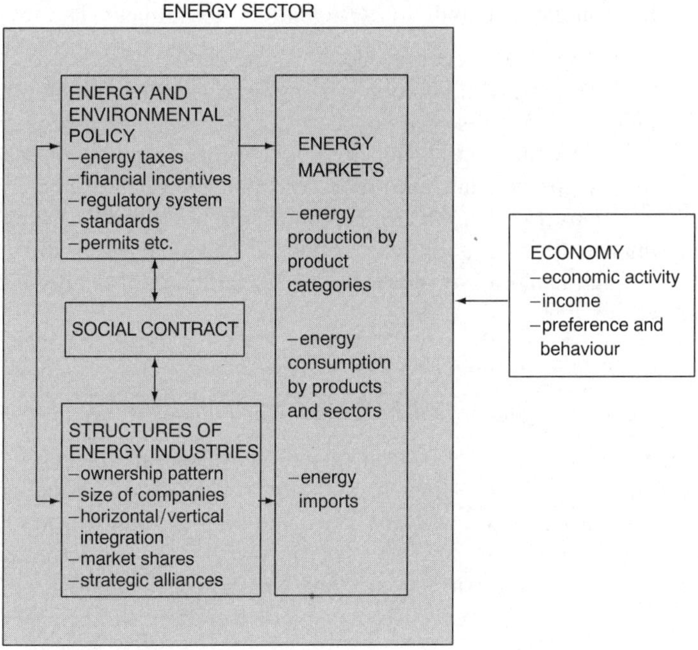

ENERGY SECTOR

FIG. 1.3. Framework of analysis

The Economy

Economic mechanisms and market forces have always been powerful in shaping the energy sector. For this reason, the quantitative part of this study starts with calculations based on economic models where the balance between supply and demand is explained by economic variables. The level and composition of energy consumption in a country will reflect the level of economic development in terms of GDP, the degree to which advanced technology has penetrated the economy, the sectoral composition in terms of economic activity, and the prices charged to end-users. Over time, changes in energy consumption will depend on the development of the same parameters. The level and composition of economic growth in conjunction with energy prices are fundamental in explaining consumption. When it comes to the relative importance of different energy carriers, again economic factors are essential, but politics from time to time outplays the market.

Politics

As underlined above, the energy sector is frequently exposed to political intervention, with varying motivation ranging from security and foreign policy to local employment and regional development. For our analytical purposes it suffices to note that such extraneous political concerns can be imposed on the energy sector by governments enforcing new policies which promote specific political objectives. In other words, under such circumstances, politics shapes policies which determine or at least influence energy markets directly (for example, through taxation) or indirectly via the structure of energy industries (see below).

Energy and Environmental Policies

Governments have a range of instruments at their disposal when they want to influence energy markets. Policy formulation can be based on energy (like security) or environmental concerns. These are often at cross purposes, as we will show below (see especially Chapter 4). At this stage we do not presume coherence or policy integration. We merely want to outline the types of policy instruments and illustrate how they can work their way through the energy sector.

It is useful to distinguish between policies that affect the framework conditions under which commercial entities have to operate (indirect action), and direct action by governments. Direct action refers to governments intervening directly in the market through state-owned companies or detailed decrees extended to commercial entities. More informal policies aimed at persuading companies to act in a particular way or voluntary agreements are also included in this category of policy measures. Complicated concession and licensing systems regulate emissions of toxic waste and all kinds of pollutants, the technology and the site for new plants, transmission lines to be built, etc. Often governments directly decide in each specific case on siting, on which fuel and what technology to use.

Among the indirect policy instruments, we shall distinguish four categories: energy taxes and subsidies, information and exhortation, technical standards, and the regulatory system in the forms of sector-specific laws and regulations. Indirect action is non-discriminatory in the sense that it has an equal effect, at least on paper, on the framework conditions of all actors in the market. As we shall see in Chapter 4, all these instruments are frequently applied to the energy sector for environmental purposes. Such policies affect energy markets, for example by influencing relative prices for end-users, and they also contribute to shaping the structure of the industries that produce energy. Governments can introduce levies or standards that favour particular energy sources, thereby setting the pattern for future production.

An example may be useful to clarify this point: The changing energy priorities of the post-war period have become embodied in capital equipment with a very long lifetime. Thus, different policies are represented in a physical way in sunk capital as distinct layers of equipment typical of its time. Take for example the power sector. In many countries there was a time for oil, later replaced by coal, followed by the more or less forced introduction of nuclear power. Gas-fired power was removed from the system, before it again became attractive from a political point of view in the mid-1990s. Bio-, solar, and wind power have also been introduced as a response to policy initiatives, not as a result of market forces. In most countries, the accumulated effect of past policies is that the park of power plants is very different from what would have come about from the free play of market forces, to the extent that an energy market devoid of political interference can be said to be anything but a theoretical construction.

This illustrates the dynamic links between the variables shaping the energy sector: Politics breeds policies that have a bearing on industrial structures, which in turn influence policies and markets. In the conclusions to Chapter 3, covering each of the main energy carriers, we will point out such links in empirical terms. Governments also shape industrial structures through legislation that protects the dominant position of specific companies, or, conversely, introduces reforms designed to break up such patterns. As we shall see below, there is now a strong tendency to direct energy policy deliberately towards altering the industrial structure, thereby setting new framework conditions for energy markets.

Structure of Energy Industries

This term refers to the composition of ownership, size of companies, and degree of vertical and horizontal integration. Size and composition in terms of large and small companies vary considerably between energy carriers and countries. For some energy carriers and for historical and political reasons, vertically integrated chains of interlocking activities under a central command have developed, while the same services in other cases may be provided by a fragmented pattern of isolated, even competing smaller units. Electricity generation and transmission of natural gas tends to be undertaken by large companies (as in Germany and the Netherlands). At the same time, distribution is undertaken by a vast number of small and locally oriented and owned companies. By contrast, in France and Italy, the same services are produced by vertically integrated country-wide government-owned companies. Private versus public ownership in energy industries is an important issue which may have important consequences for energy trends. Therefore, ownership across energy products and across countries will be described and analysed.

Structural differences often have important bearings on how the sector responds to new challenges or opportunities. One example may illustrate this: a highly centralized structure dominated by large companies tends to favour large-scale technological solutions involving high-risk R&D components. A decentralized system with a large number of commercial entities would opt instead for small-scale, low-risk technologies.

Industrial structures influence energy markets by shaping the supply of energy products, but they also play an indirect role through interaction with government policies. Energy industries are not mere passive recipients of political orders from above. It is a basic assumption in this study that they interact closely with public decision-makers in order to influence policy conditions in a manner compatible with their interests as corporations.

Thus, the linkage between the governmental and the industrial level works both ways, as indicated in Fig. 1.3. In some cases, this interaction is so tight that it is hard to tell where the public sector ends and the corporate world begins. This is particularly tricky when large state-owned corporations are involved in policy-making, as they can be both formulators and executors of public policy. However, for analytical purposes we maintain a distinction between energy and environmental policies, as decided by public bodies, on the one hand, and industrial structure on the other. The critical point is the link between them, which we have conceptualized in the term 'social contract'.

The Social Contract

For the reasons given above, energy is seen as too important to be relegated to the market alone. All over Europe, governments have kept and keep a close watch on energy supplies, as politicians know that they will ultimately be held accountable if this lifeline of modern society is disturbed. They have consequently developed close, but diverse links with the suppliers of energy. In some countries the public interest is handled through direct ownership in major corporations. In others, the state confines itself to defining the framework conditions for energy markets through legislation, taxation, or other means of regulation. This indirect influence can be just as effective in promoting governmental objectives as direct participation.

Whatever the means, the essence is the same: there must be a fundamental common understanding between the energy industries and society at large, represented by governments. This is the crux of the social contract that binds together the policy and industrial levels described above, by defining the rights and obligations of the two sides. Through this 'quid pro quo', actors in the sector are given certain societal objectives to which they are committed— security of supply, employment, environmental goals, etc. In return, they acquire a carefully defined freedom of action. This can take the form of legal or quasi-legal monopoly rights, for example within a specific area. Governments can favour one source of energy at the expense of another, through taxation in return for regional development, employment, or some other laudable goal. An energy company can in addition be given access to the treasury by way of state aid or hidden subsidies.

The social contract can assume different forms, ranging from a formal obligation written into the statutes of a public enterprise to a purely informal understanding—a tacit bargain—between government officials and industry executives. The former is, needless to say, more resilient to changing political winds than the latter. A social contract is never perfectly stable, as it defines not only winners, but also losers, who will try to exploit new circumstances to their advantage. As we shall see below, the 1990s offer ample opportunities in this respect.

The durability of a social contract will depend on the ability of the energy industries involved to deliver the goods that are expected or required from them—whether it be employment, national pride, technological sophistication, or environmental clean-up. If they deliver, the corporations in question have kept their part of the bargain, and can expect governments to honour the other side in terms of legal protection, state aids, tax favours, and the like.

1.4. Study Outline

In Part I we undertake a twenty-five-year retrospective analysis of the energy sector in Western and Central Europe. We describe and analyse the structure of the energy industries, energy policies, and markets, and the relationship between them (Chapters 2–6). For each subsector we summarize the tensions between the established structures and the challenges from the new agenda outlined above. Thereafter we review the major environmental problems emanating from the energy sector and the main institutional responses from governments, including efforts towards cooperation at the European level (Chapters 7 and 8).

In Chapter 9 we outline the major elements of our scenarios for the future, with a focus on political and economic integration in Europe and the strength of environmental concern. In Chapters 10 and 11 we show through quantitative calculations how energy sector and markets are likely to develop under two qualitatively different paths in political/economic development in Europe, while environmental concern remains at today's level. In Chapter 12 we present a systematic comparison between the two political/economic scenarios, demonstrating their impact on energy consumption, fuel mix, and emissions of pollutants. In Chapter 13 we return to the fundamental environmental challenge: what if climate change becomes a dominant political force? This question is answered at the end of the book by imposing a much higher level of environmental concern on our political/economic scenarios. This approach follows from our framework of analysis as set out above: whatever strength the politics of climate change will have in the future, it will have to grapple with the intricacies of the real world. While comprehensive and far-reaching, the agenda set by climate change does not provide a blueprint for restructuring an energy sector entrenched in a past with very different priorities. How such a transformation will come about, if it does, depends in a critical sense on the structural make-up of the energy sector: on the influence of politics, on the relations among key actors and the weight of competing energy products. This is the essence of the conclusions to this book, as presented in Chapter 13.

PART I

Trends and Policies
1970–1995

2
Energy Demand: Trends and Policies

Energy consumption grew steadily at a high annual rate until 1973, but growth rates have fluctuated since the first oil-price shock. Moreover, the growth pattern has differed across European countries, with high growth rates in South and Central Europe and more modest increases in the North. Some of these differences are explained by different economic growth rates. Variations in energy price trends, partly due to price regulations and taxation policies, are also important determinants. However, these are not the only explanatory factors for patterns of energy demand. The composition of economic activity is important. Economic activities have different energy requirements and changes in industry structures thus impact energy demand. Growth in income will also lead to saturation for some consumer goods and spur changes in lifestyles, some requiring more energy, others less. Some of these structural changes are advanced by energy price changes, but to a large extent they take place irrespective of price movements. Finally, demand-side policies, primarily through non-pricing energy conservation efforts, have affected trends in energy consumption.

This chapter reviews the main trends in energy demand and discusses its determinants. Understanding how various factors have influenced past trends is crucial in order to analyse future energy and environmental trends. In this respect it is important to examine the relative impacts of 'market forces' and governmental policies on energy consumption and energy efficiency trends.

2.1. Trends in Energy Demand

Determinants for Energy Demand

Most energy analysts today agree that energy at the aggregate level is primarily driven by economic activity, income, and prices—both energy prices and prices of substitutes. These key parameters are generally considered to be important factors that influence the decisions and behaviour that affect energy demand. Prices and income also have an impact on the efficiency with which energy is used and on changes in industry structure. For example, the decline of energy-intensive industries in favour of light industry, or the less energy-consuming service sector, will reduce energy demand. There are also structural changes in other sectors, such as in transport. The importance of rail transport has continuously declined since 1970, and been more than compensated for by increased air and road transport. These modal shifts contributed to a strong growth in overall energy demand.

Aggregate energy demand is also influenced by the fuel mix. The efficiency with which energy is converted into useful energy or service depends on the technology and fuel used. The replacement of coal in industry by gas or oil, or oil in the residential sector (households) by gas or electricity has an impact on energy demand. Fuel substitution may increase or decrease overall demand. The use of gas in industrial processes normally leads to a higher conversion efficiency. But in the residential sector, the replacement of coal- or oil-fired installations by gas or electricity may lead to higher energy consumption through an increase in the service level, as single furnace systems that normally heat only a selected number of rooms are replaced by central heating, providing warmth for the entire home.

Energy demand is also influenced by policy. As we shall argue in this book, policy mainly affects energy demand through price regulations and taxation, and through the direct regulation of the fuel mix in, for example, the electricity sector. Still most governments have in place (non-price) programmes to encourage energy efficiency. The policy objectives of these programmes have shifted over time, but a persistent motive for policy interventions has been the perception that market imperfections and barriers inhibit cost-efficient use of energy. Such barriers include lack of information and technical skills, restrictions on access to capital, and budgetary constraints.

Energy Demand in Western Europe

Main Trends

Energy consumption in Western Europe grew at an average annual rate of 4.5 per cent in the first years of the 1970s. This was a slow-down compared to the strong growth of 6.5 per cent recorded in the 1960s. The first interruption in energy consumption growth recorded for the post-war period came as a result of the oil-price shock in 1973. From 1973 to 1975 the consumption level fell by 6 per cent, but growth recovered and stayed at an annual rate of 3.6 per cent from 1975 to 1979. After the second oil crisis energy consumption started a

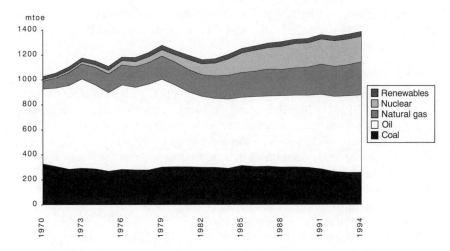

FIG. 2.1. Energy consumption in Western Europe by fuel
Source: IEA (1996)

three-year decline which compounded to a 10 per cent reduction in the consumption level from 1979 to 1982. The 1979 level was not regained before 1986, after which consumption has continued to grow, albeit at a fairly modest annual rate of 1.5 per cent. High energy prices resulting from the oil crisis, reductions in economic output during the two oil crises and the impact of the recession which began to hit Europe in 1991, and eventually saturation in some categories of energy use are the main reasons why the high growth of the 1960s has not been repeated since.

Changes in the Fuel Mix

The contraction of coal demand which started in the mid-1950s continued during the first part of the 1970s. From 1974 the consumption level stabilized as higher demand from the electricity sector compensated for the continued loss of market share in industry and the residential sector. Oil, which had grown rapidly until the oil-price shock in 1973, saw its share of total primary energy supply (TPES) fall from 60 per cent in 1973 to less than 45 per cent in 1994. Oil consumption fell by 20 per cent from 1979 to 1983. It recuperated somewhat after 1986 but was still in 1994 some 15 per cent below its peak level in 1973. Natural gas and nuclear power have taken large new market shares from 1970 to 1994, natural gas primarily during the first part of the 1970s when gas use increased at an average rate of 19 per cent. Nuclear power grew most rapidly after 1980 as new power plants came on stream in

France, Belgium, and Germany. Trends in supply and demand of the various
energy forms and the prices influencing these trends are described in greater
detail in Chapters 3–6.

Energy Intensity

For the whole of Western Europe energy consumption grew by a modest 33 per
cent from 1970 to 1994. GDP grew by 75 per cent over the same period.
Energy consumption growth rates have been lower than GDP growth for nearly
every year since 1973. This is reflected in the development of the energy
intensity (i.e. the energy requirement of the entire economy per unit of GDP)
which has dropped by 23 per cent from 1970 to 1993 (see Fig. 2.2). Even
though trends in energy intensity are not an accurate measure of energy
efficiency (see Box 2.1) this development suggests that energy efficiency in
Europe has improved markedly since 1970.

The largest declines were experienced in the years following the second oil-
price shocks—between 1980 and 1983 energy intensity declined by 2.4 per
cent per annum and between 1987 and 1990 by 2.8 per cent per annum. In later
years, energy intensity decreased only modestly, which is an indication that
lower energy prices have reduced the incentive for energy efficiency.

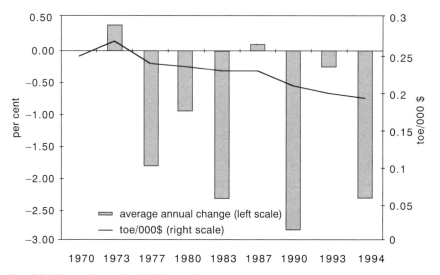

Fɪɢ. 2.2. Energy intensity in Western Europe

Box 2.1 *Energy intensity and energy efficiency*

Energy intensities are often used as indicators of energy efficiency. Energy intensity expresses the energy consumption per unit of activity or some other variable related to energy consumption. At the national level, energy consumption per unit of economic output (GDP) is often used as an indicator of energy efficiency performance. At more disaggregated levels, energy intensities are often expressed in physical units, e.g. energy consumption per tonne of steel production and energy use for space heating per m^2.

Energy efficiency is a technical term denoting the output of services in relation to energy input, such as petrol consumption of a car travelling a certain distance (litres/100 km). The quantification of energy efficiency requires a very detailed assessment of the efficiency of the technology in place. Efficiency data are therefore scarce. Moreover, since they are expressed in different physical units they cannot easily be aggregated.

Energy intensity is, however, only a proxy for energy efficiency. Efficiency and intensity may show diverging trends for three reasons:

Structural change. The energy efficiency of different sectors varies. Since energy consumption growth also differs by sector, aggregate energy intensity (e.g. energy consumption per GDP) will change even with constant efficiencies across sectors. Industrial activities generally require more energy per unit of value added than production in agriculture, commerce, and public services. Relative decline in industrial production will therefore tend to lower the aggregate energy intensity. Saturation effects (e.g. of electrical appliances in households) also play a major role for energy consumption trends and may contribute to lowering the energy intensities.

Lifestyle and behavioural changes. Changes in lifestyle, for example flexible working hours or reduced family size, may increase or decrease energy demand. Furthermore, energy consumers can decide to lessen the demand for energy services (save energy), without this necessarily affecting economic activity or other parameters being used in energy intensity calculations. In this case energy intensity will fall with constant energy efficiency: e.g. space heating requirements through reduction in indoor temperature.

> *Fuel substitution*. If energy intensity is calculated on the basis of the aggregation of different fuels, fuel substitution, e.g. the switch from oil to electricity for space heating, may increase the energy intensity of the economy, if the conversion efficiency in power plants is less than the efficiency of an oil-fired heating system.

Sectoral Trends

Exploring trends in energy consumption and efficiency requires a split into the three main sectors of final energy consumption:

- Industry, encompassing manufacturing, mining, and construction. In the first part of the 1990s coal, natural gas, and electricity each have a 20–5 per cent share of the industrial market, while oil products account for about 30 per cent.
- Transport, which includes road, railway, air, coastal shipping, and fisheries. Road transport is the most important, with 85 per cent, followed by air, 10 per cent. Some 98 per cent of transport consumption is of oil products.
- Other sectors comprising agriculture, residential, commercial, and public services. The residential sector is the largest within this category, with two-thirds of consumption. Commercial and public services together represent 27 per cent of final energy consumption.

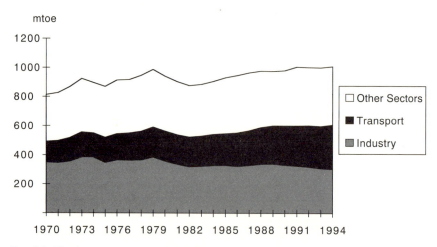

FIG. 2.3. Final energy consumption in Western Europe by sector
Source: IEA (1994*c*)

Industry

Industry has for most of the period since 1970 been the principal energy-consuming sector. In 1970 industrial energy use represented close to one half of total final consumption. Consumption increased only marginally during the 1970s and fell by 20 per cent by 1982 after the second oil-price shock. During most of the 1980s, industrial energy consumption has remained relatively stable within a band of $+/- 4$ per cent. The consumption level in 1994 was some 5 per cent below the 1970 level, despite a 50 per cent increase in value added.

Some countries have had developments significantly different from this general trend. Southern European countries with relatively low GDP per capita have experienced rapid growth in industrial production since the early 1970s, with an ensuing high growth in industrial energy use. The larger continental countries, Germany, France, and Italy, had in 1994 industrial energy consumption close to the 1970 level. This contrasts sharply with a 30 per cent drop in the case of the United Kingdom.

The increase in industrial value added, coupled with a stagnant consumption trend, signifies large declines in energy intensity. Intensity has dropped nearly every year since 1970 (see Fig. 2.4). It is interesting to note that the decline in industrial energy intensity was well under way before the first oil-price shock. From 1970 to 1973 the intensity fell annually by 1.2 per cent, only marginally lower than the decline registered for the rest of the 1970s. In the first part of the 1980s the decline accelerated to more than 2 per cent as energy prices spurred technical efficiency improvements and also a contraction of energy-intensive industries and relocation of production to countries outside Western Europe, in cases such as steel and ethylene production (IEA, 1994*c*). Energy prices were not the only reason for the industry to settle outside developed countries. Cheap access to raw. materials and generally less stringent environmental regulation were also conducive factors.

Energy intensity in industry continued to decrease as energy prices started to decline in 1986. This is explained by further restructuring towards less energy-intensive and higher value added industries and the impact of large investment and turnover of the capital stock, thus increasing the market penetration of modern and generally more efficient technology. For example, in the steel industry of Western Europe energy use per tonne of crude steel declined by 15 per cent between 1980 and 1988. The halt in the intensity reduction that set in after 1990 was primarily a result of a large drop in industrial investments caused by the economic recession.

This description illustrates that the decline in industrial energy intensity can be attributed to two elements: (i) a structural change in industrial production

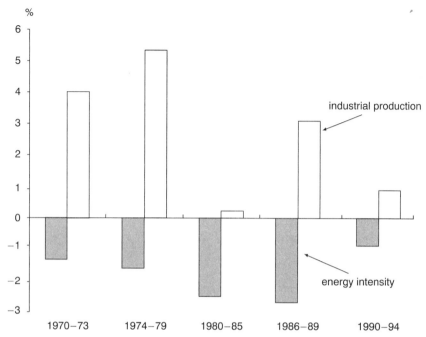

Fɪɢ. 2.4. Annual average growth in industrial production and reduction in energy intensity in Western Europe
Source: IEA (1994c) and OECD (1994)

towards less energy-intensive products and relocation of energy intensive industries to countries outside Europe; and (ii) energy-efficiency improvements in industrial processes. In most European countries, structural changes contributed to a reduction of energy intensity. A study on the developments in industrial energy demand in the 1980s for major European countries found that only in France did structural changes lead to a higher energy intensity (Schipper and Meyers, 1992). The main reason for the continued dominance of energy-intensive industries in France is the low electricity prices offered to large industrial customers. In Germany about 30 per cent of the total reduction in energy intensity resulted from a sectoral shift, closing down production units in iron, steel, and non-ferrous metals. Unlike France, Germany had a substantial increase in industrial energy prices. This probably was a major factor behind the change in industry structure. Intensity reductions were substantially less in Western Europe than in the USA or Japan. One reason for this is the slow growth in industrial production in Western Europe. Low growth delays structural change and efficiency improvements through investments in new technologies.

These developments illustrate also the impact of broader policy objectives on energy demand. Germany and the United Kingdom have pursued a less protective industrial policy and phased out large parts of their energy-intensive industries. France has been slower in industrial restructuring, partly due to excess capacity of electricity which allowed French industry to be supplied with relatively cheap electricity, but also due to social and industrial policy objectives.

Environmental policies have also played a role in some countries. For example, the paper and pulp industry has significant environmental impact as large amounts of water and energy are used. In the Nordic countries and Austria, for example, stringent environmental legislation, linked to the quality of discharged water or requirements for recycling, led to the introduction of less energy-consuming production processes. Examples are the use of certain by-products for heat production which would otherwise be disposed of in the water, or the use of recycled waste paper which requires less energy to reform into paper than does virgin wood.

Transport

Transport fuel consumption has increased every year since 1970 except for 1974 when there were physical demand restraints on oil in response to the first oil crisis. This has increased more than any other component of energy consumption, and its share of total consumption has grown from 20 per cent in 1970 to 30 per cent in 1994. From 1970 to 1973 the average annual growth rate was 6.4 per cent, from 1975 to 1979 4.4 per cent, and from 1986 to 1990, when the increase was spurred by a drop in oil prices, the annual growth rate was 4.2 per cent. Since 1990 transport fuel consumption has decelerated to 2.0 per cent.

The dominant share of transport fuels is accounted for by road transport (85 per cent). Air transport, however, has increased steeply since 1970 and now has a 12 per cent of total transport fuel use. Among the major Western European countries there has been little difference in the growth in the use of transport fuels. France, Italy, and Germany have all seen a doubling in consumption from 1970 to 1994. United Kingdom, the only major European country with a slightly different trend, has had an increase of 70 per cent.

The volumes of transport, i.e. passenger km and goods km, both increased by more than 100 per cent between 1970 and 1994, i.e. more than GDP and private disposable income which grew by about 75 per cent. The growth in transport fuel consumption and car ownership has been relatively stable despite periods of sharp fuel price increases and years of stagnation or slow growth in GDP and income. From 1970 to 1988 the road vehicle stock more than doubled, with slightly higher growth for passenger cars than freight vehicles (IEA, 1991*a*). The large increase in the number of cars has been the major factor behind the increase in transport fuel consumption. In addition two other factors influence consumption:

- Distance travelled per car has to a small extent offset this increase in car ownership as the average distance travelled per passenger car fell by 5–10 per cent in most West European countries between 1970 and 1980. Since 1985, however, average distance travelled has been growing again (Schipper *et al.*, 1993).
- Increased average fuel efficiency (actual fuel use per km driven, calculated as an average of the total car fleet on the road) has only marginally contributed to a lower growth in fuel consumption. Comparable fuel efficiency data do not exist for all countries, but data for Germany, France, Norway, and Sweden for the years 1970 to 1990 indicate that fuel use per km has changed little (Schipper *et al.*, 1993).

The stable average fuel efficiency may seem surprising in view of the considerable achievements in technical efficiency for most car models. All major car manufacturers currently have cars on the market with a specific consumption of 4.4 litres/100 km (official consumption tests), see Grubb and Walker (1992), whereas the average fuel use has remained at about 10 litres/100 km since the late 1970s.

The main reason for the high average fuel requirements is the increase of engine capacity and weight of new cars in Western Europe. In Western Germany, for example, the share of cars with engine capacity under 1,500cc declined from 50 per cent in 1978 to 39 per cent in 1988. This decrease was compensated by an increase from 38 per cent to 55 per cent of the share of larger, more powerful, and fuel-consuming cars in the 1,500 to 2,000cc range. Similar trends, which are clearly a result of growth in disposable income, have been observed for other Western European countries, see IEA (1991*a*).

But the purchase of larger cars is not the only factor behind the apparent lack of efficiency improvements in road transport. There are three additional reasons why improvements in fuel efficiency of new cars have not translated into improved efficiency of the average car fleet:

- Changes in the pattern of vehicle use towards urban driving and driving on congested roads lead to higher fuel consumption than indicated in the official test consumption.

- More powerful vehicles and better roads lead to higher driving speed and more fuel use per km.
- Stringent emission standards have also contributed to increased fuel use per km. Three-way catalytic converters which are now mandatory in most West European countries increase fuel use by 5–10 per cent.

As in other sectors, changes in the price of transport fuel influence energy demand. In response to increases, consumers tend to use their car less. Before 1973, relatively low prices for automotive fuel and increased market penetration by private cars resulted in a considerable annual growth of oil consumption in the transport sector. Expenditure on transport fuel was not an issue to be considered when cars were bought or used. Between 1970 and 1975 oil demand for transport increased by 6.4 per cent annually.

After the first and second oil-price shock, the automobile industry and consumers both responded to these price hikes. Car manufacturers started to develop and produce more energy-efficient cars. These cars began to penetrate the market in the second half of the 1970s. Furthermore, consumers cut down on the use of their cars, resulting in a decline in the average distance travelled (IEA, 1991*a*). The annual growth rates dropped to 2.4 per cent for the years 1975 to 1985. Since 1985, relatively stable transport fuel prices have led to an increase in the average distance travelled, which—together with a levelling off of improvements in the efficiency of new cars—resulted in an increase of annual growth rates to 3.5 per cent for the years 1985 to 1994.

Governments in Europe have used taxes on transport fuel primarily for budgetary reasons, but also considering security of supply, given the sector's dependence on oil. The share of taxes in transport fuel—in particular for gasoline—is generally the highest of all end-use prices. In Europe, for example, the tax share in unleaded gasoline is substantially above the actual production costs, ranging from 50 to 75 per cent of the end-user price. Transport fuel is also the main target for the introduction of environmental taxes. Gasoline taxes have influenced energy demand. Not surprisingly, in the United States, where gasoline taxes are the lowest in the OECD, the average fuel consumption ranks among the highest.

The prevailing trends in energy demand in transport since the 1970s highlight the continuing significant growth potential of the transport sector, and suggest that car ownership has not yet reached saturation level. Efficiency improvements have not had a significant impact on consumption in the road transport segment. Changes in behaviour and lifestyle, such as the preference for more powerful cars and weekend travel, have offset these efficiency improvements. As disposable income grew in the 1980s, consumers tended to invest in more powerful cars and higher fuel efficiency (entailing a lower fuel bill) has been a second-order consideration. Given the important role that mobility seems to play in our society, the trend towards growing transport fuel consumption cannot easily be reversed.

Other Sectors

Energy demand in the residential, commercial/public services, and agricultural sectors accounts for about 40 per cent of total energy consumption in Western Europe. Residential energy consumption has increased by one-third from 1970 to 1993, i.e. in line with the trend of total final consumption. The determinants of energy demand in the residential sector, which accounts for 58 per cent of the demand in this category, will be discussed in more detail below.

Between 1970 and 1994 the service sector was the second fastest growing energy end-use sector after transport. This is explained by the strong economic performance of commercial services requiring additional office space, which is a main determinant for energy demand. Modern office buildings generally require sophisticated heating, ventilation, and air-conditioning systems. Consequently, space conditioning (heating and cooling) makes up approximately 60 per cent of the sector's energy demand (IEA, 1991b). Lighting and electric appliances account for another 29 per cent of the consumption.

Energy consumption in households is dominated by space conditioning and water heating. Together these two categories account for 80 per cent of residential energy use (see Fig. 2.5). Changes in lifestyle have a significant impact on energy demand in the residential sector. The trend towards smaller family sizes, the market penetration of residential appliances which reduce time spent on housework, thus increasing leisure time, and the introduction of new technologies, for example in communications and computerization, generally lead to a more intensive use of energy.

Population growth and the number of persons living in a household (household size) are important determinants for residential energy use. From 1970 to 1994 the population grew by 15 per cent in Western Europe. At the same time household size fell by about 15 per cent in the major West European countries. Hence residential energy consumption per capita increased by 17 per cent from 1970 to 1994 but the consumption per household remained almost unchanged.

The growth in consumption caused by growth in the number of households and disposable income has partly been offset by energy-efficiency improvements. Energy consumption for space conditioning per m^2 fell by 15 per cent from 1973 to 1988 in the four major European countries (West Germany, France, United Kingdom, and Italy) and by 24 per cent in the Scandinavian countries (Schipper and Meyers, 1992). The improvements were caused by

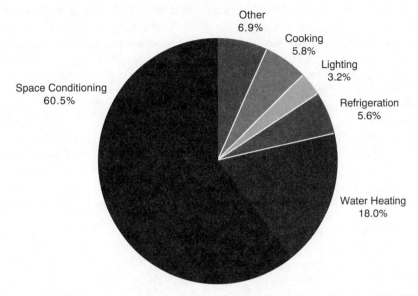

FIG. 2.5. Residential energy consumption by end-use category in the IEA countries, 1988
Source: IEA (1991*b*)

reductions in indoor temperature, improved insulation, and higher efficiency of heating installations. Surveys in Denmark and Germany show sharp declines in indoor temperature in 1973–5 and 1979–80, following steep increases in household energy prices. A gradual rebound in the temperature followed as end-use prices stabilized and fell.

Higher thermal efficiency in new houses, which is, for example, triggered by stringent building codes, has played a relatively modest role, since the growth in the stock of houses has been moderate. Of the 1988 stock of private homes in Western Europe only 15–18 per cent were built after 1974.

Consumption of electricity by appliances has had the greatest impact on residential energy demand, with an increase between 1972 and 1988 of 125 per cent for four major European countries, France, Germany, Italy, and the United Kingdom, and 59 per cent in Scandinavia (Schipper and Meyers, 1992). Market penetration of refrigerators, freezers, washing machines, and dish washers has been the driving force and has more than offset the efficiency improvements since 1970. By 1988 refrigerators and washing machines in Western Europe were close to saturation, with more than 100 and 90 units per 100 households, respectively. However, market penetration of freezers was far from saturation (IEA, 1989*a*). Moreover, new appliances, like personal computers, have begun to enter the market, resulting in an additional impact on the growth of energy demand, mostly for electricity. Similar to the impact on energy demand for space conditioning, the trend towards a smaller household size, which leads to a growing number of apartments, will also contribute to a continuous growth in residential energy demand.

Nevertheless, the scope for improvement in the efficiency of appliances is substantial. Often the best available technologies have specific energy requirements which are half those of the average consumption in the existing stock (IEA, 1991*b*). Efficiency improvements will therefore gain momentum as a large number of appliances are replaced by new and more efficient models. Though these developments will undoubtedly reduce the pressure on growth, it is unlikely to contribute to a reduction in energy demand, as the number of appliances will continue to increase and thus outpace efficiency improvements.

Energy Demand in Central Europe

Main Trends

Growth in energy consumption in Central Europe since 1970 has been higher than in Western Europe. During the 1970s the annual growth was steady at 4 per cent. Central European countries were also affected by the second oil-price shock, albeit to a lesser degree than those in Western Europe. In the 1970s the Soviet Union—the single oil supplier for import dependent satellite states—pursued the policy of shielding these states from the impact of accelerating energy prices. In the early 1980s, the Soviet Union changed this policy and took steps to bring oil export prices to their CMEA partners more in line with world-market prices. The global economic recession in the early 1980s also had its impact on economic activity in Central Europe. Energy consumption fell slightly from 1980 to 1981 and subsequently remained stagnant to 1984. Consumption grew again from 1984 to 1987 but at a lower rate than during the 1970s reflecting weaker economic growth. Energy consumption in Central Europe peaked in 1987 at 50 per cent above the 1971 consumption level. Over the same period consumption growth in Western Europe was 25 per cent.

Energy demand started to change dramatically towards the end of the 1980s. The collapse of the Communist system resulted in an unprecedented decline in the economic output of all Central European countries and caused a substantial contraction in energy demand. Between 1989 and 1993 GDP declined in the Slovak Republic by 25 per cent, in Poland by 14 per cent, and in Hungary by more than 20 per cent. Though the share of coal in primary energy supplies remained high, coal has taken the bulk of the reduction in consumption since

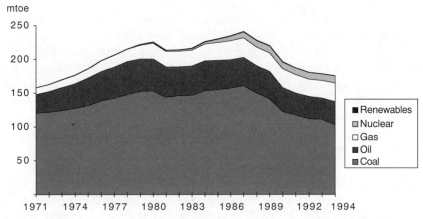

FIG. 2.6. Energy consumption by primary energy products in Central Europe

the late 1980s. This is explained by the collapse of energy-intensive industry and contraction of electricity generation in which coal was the main fuel input.

Energy Intensities

There are only limited data available on Central European energy intensities. It is however clear that they are considerably higher than in Western Europe (see Fig. 2.7). This is partly caused by energy-intensive industries taking a large share of total manufacturing and inefficient production. Furthermore, it is

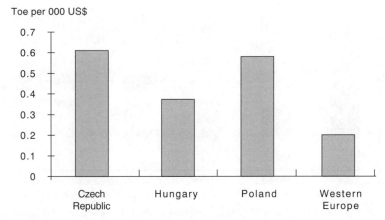

FIG. 2.7. Energy intensity for Central Europe, 1993

unlikely that the calculation of GDP fully captures the true value of economic output. The contribution of the black economy is considered significant, but is not fully accounted for in GDP.

Unlike in the West, energy intensities in Central Europe grew until 1987. This was a result of the shielding against price increases (the absence of a budget constraint, as Central European industries obtained energy via state allocation rather than purchase), industrialization, and growth of energy-intensive industries.

Sectoral Trends

Developments in energy consumption in the main energy-consuming sectors, industry, transport, and households, were substantially different to those observed in Western Europe. Prior to the collapse of the old regimes, final consumption in Central Europe was characterized by predominant energy demand in industry. Contrary to Western Europe, energy use for transport is low. Though the differences between East and West in the composition of total final consumption have narrowed in recent years, they are still substantial, as Fig. 2.8 illustrates. In 1993 the share of industry was eight percentage points higher in Central Europe, but the energy demand in the transport sector was sixteen percentage points below the Western European average.

Industry

The high concentration of energy consumption in industry is explained by the importance of the industrial sector, and in particular of energy-intensive

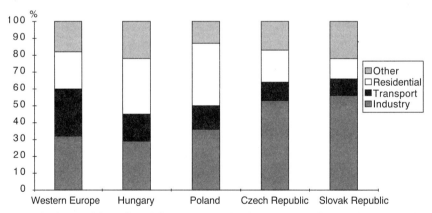

Fɪɢ. 2.8. Composition of total final consumption in Western and Central Europe, 1992
Source: IEA (1994c)

industries. Manufacturing industry has used outdated machinery and equipment and exhibited lack of maintenance and repair, managerial deficiencies,
and poor operational practices due to lack of proper incentives to pursue
economic efficiency. For example, in 1988 over 80 per cent of Poland's capital
stock in industry was over five years old, compared to 40 per cent in Germany
(IEA, 1995*a*), and the Czech steel industry requires at least 35 per cent more
energy to produce a ton of pig iron than in Western Europe (IEA, 1994*b*).

Between 1989 and 1993 industrial production in all Central European
countries declined by more than 25 per cent, and by almost 40 per cent in
the Slovak Republic (EBRD, 1994). These contractions were caused by the
collapse of trade among traditional partners, the contraction of domestic
demand, and the absence of new export markets. Though the down-turns began
to level off in 1992—Poland and Hungary are already experiencing an
upswing in industrial output—old patterns of industrial energy use will not
re-emerge. Industrial recovery is driven by the settlement of new and higher
value-added industries, by investment in capital stock thus replacing old and
outdated technology, and closure of unproductive industrial activities.

The decline in industrial output has not always resulted in decreased energy
intensity. While the decline in output and structural shifts in Poland have been
accompanied by a reduction in energy use, intensity has increased by 7.6 per
cent between 1989 and 1992, reflecting a deterioration in efficiency, as in
many cases planned capacity has been under-utilized (IEA, 1995*a*). In the
early years of transition towards market economies, plant managers may have
been more preoccupied with more immediate questions of survival and marketing, rather than energy efficiency. Furthermore, energy prices are still
below market levels and some state enterprises continue to operate under a
soft budget constraint which allows them to accumulate unpaid energy bills
and to subsidize production. Moreover, some legislation essential for accountable and commercial business operation, such as bankruptcy and company
laws, has only recently been implemented.

The industrial structure of Central Europe points at a significant potential for
further efficiency improvements. Available data suggest that over the next
seven to ten years energy savings in Czech industry could reach a total of about
20–5 per cent of current levels (IEA, 1994*b*). But there is some uncertainty
regarding the extent to which efficiency improvements can contribute to overall
demand reductions and how sustainable they will be. Some industries with high
efficiency potential, such as iron, steel, or chemicals, may not survive in the
medium term. Moreover, in order to achieve such savings, it would be necessary for energy prices to continue to rise and for recession to be overcome.
Without economic recovery, industry would not be able to modernize its capital

stock, but this would be a prerequisite for the introduction of modern and energy efficient technologies. Economic recovery would also be necessary to allow investment in higher value-added industries, thus facilitating further industrial restructuring, which may have a larger impact on energy demand than increased efficiency.

Transport

Energy demand in the transport sector was in 1994 only 12 per cent of total final consumption, compared to almost 30 per cent in Western Europe. Also the modal split is significantly different: the share of air transport is generally lower, whereas rail services are still important contributors to passenger and freight transport. Since the break-up of the Communist system, however, the importance of rail transport has begun to decrease. Furthermore, the structure of vehicle ownership has also begun to change: in the Communist system, private car ownership was restricted either by law or through a physical shortage of cars, long waiting lists, and exorbitantly high prices. Consequently, energy demand for private transport was limited and just a minor fraction of the transport sector's consumption.

The collapse of the Communist system had a significant impact on the transport sector. At first, there was excessive demand for private cars which could not be met domestically. As a result, many old and inefficient cars were imported into Central Europe. In light of these negative developments, which boosted oil demand and also jeopardized the environment and safety, some countries began to restrict imports of used cars. Regulations have been introduced in Hungary and Poland to limit the maximum age of imported cars, trucks, and buses to six years.

Although the market penetration of private cars has increased since 1989, there is still a substantial gap compared to the West. In 1992 in Hungary car ownership was only 0.2 per capita whereas the figure was 0.42 in France and Austria. The commercial transport fleet is also very small but increasing rapidly, replacing rail as preferred transport mode for freight. Available data indicate that the efficiency of the average car fleet in Central Europe is not much different from Western European countries. For example, Polish automobiles in 1989 appear to require approximately 8–9 litres/100 km, which is similar to efficiency levels observed in Germany and Sweden.

The recession following the collapse of the Communist regime, the decrease in real income, budget constraints, and petrol prices approaching market levels have clearly restricted mobility and reduced the pressure on energy demand in the sector. In Hungary annual car use is currently estimated to be 7,000–8,000 km, compared to 12,000–15,000 km in Western Europe and 9,000–10,000 km in Hungary in the 1980s, leading to a reduction in energy demand for private transport. This trend was not uniform across Central Europe: in the Czech Republic consumption of transport fuels increased by 30 per cent between 1991 and 1994, which is possibly a reflection of relatively high income. The per capita income is about 40 per cent higher in the Czech Republic than in Hungary.

Central European countries are a preferred target for Western car manufacturers for green-field projects or the establishment of joint-ventures with domestic producers, thus making available modern technology. As production will be targeted for export markets, efficiency will approach those of similar sized cars produced in the West. In light of prevailing income limitations, which will continue in the medium term, smaller cars are likely to be the preferred choice and mobility will continue to be a service more than luxury as in the West. Lifestyle changes—calling for higher mobility—will be a driving force for transport energy demand in Central Europe. Once the current recession is overcome, it is likely that the transport sector will undergo growth even in excess of that experienced in the West in the 1970s.

Residential Sector

Although accurate statistics are lacking in Central Europe there is evidence that residential energy consumption has increased markedly since 1970. As in the West, energy for space conditioning makes up the largest share of energy demand in households. In Poland, 70 to 80 per cent of residential energy consumption is for space and water heating, in Hungary the share is 69 per cent, leaving the remainder for electric appliances, including lighting and cooking.

End-use efficiency has changed little in consideration of the large increase in floor space. From 1980 to 1989 the floor space in the residential sector in Hungary increased by close to 40 per cent while energy consumption per m^2 only dropped by 10 per cent. Though energy demand per household is generally lower than in Western Europe, energy consumption per m^2 is above or similar to the levels observed in the West.

Central European countries exhibit a large share of households connected to district heating systems: one-third of the households in the Czech Republic and 16 per cent in Hungary, for example. In Poland, 80 per cent of the apartments in urban areas are supplied by district heating. If the transmission and distribution is well maintained, such systems are usually an efficient system for heat and hot water supply. However, distribution networks in Central European countries show losses of up to 30 per cent, due to poor quality and lack of maintenance. Furthermore, single-pipe distribution systems have been widely used where radiators have no bypass piping and cannot be shut off without cutting the flow of hot water to other radiators in the building block supplied by the single pipe. As a result, the inhabitants cannot regulate the indoor temperature and save energy by reducing the water flow. Furthermore, the quality of insulation of apartment blocks is generally low, primarily due to the widespread use of prefabricated concrete panels. Another deficiency of district heating systems is that heat meters are uncommon and consumers often pay for heat by a flat rate per m^2 of space. These deficiencies are now gradually being eliminated. For example, in the Czech Republic heat meters were installed at the entrance to all buildings in 1993.

After the fall of the Communist regimes and the transition to market economies, energy prices in Central European countries increased. This was partially a result of spiralling inflation during the early years of transition and partially caused by a gradual liberalization of the price regime. However, energy prices did not increase in real terms in all countries or for all end-use sectors. In Hungary, for example, between 1989 and 1994 electricity and gas prices for households increased by about 280 per cent in nominal terms, whereas the consumer price index and wages in industry increased by about 310 per cent. Between 1989 and 1993 in the Slovak Republic inflation exceeded the nominal increase of energy prices for households. These prices, however, grew faster than average wages. In Poland, energy prices for households grew about ten times faster than wages or inflation.

Radical price increase will be required to provide the energy industry with sufficient capital to undertake the investment that is required to reorient and restructure the energy sector. In most Central European countries the initial steps towards price liberalization will not be sufficient. Though there are some exceptions, such as transport fuel prices which are quickly approaching market levels, including tax shares similar to the West, most energy prices will need to be raised substantially. Such increases are particularly required for prices of grid-bound energies, such as electricity and district heat, where large investments are required for production and transmission. Though demand for electricity, gas, and heat is currently depressed as a result of the collapse of manufacturing industry, energy producers will now need to modernize their infrastructure in order to provide sufficient and reliable supplies once the depression is overcome. Real price increases for electricity or district heat by a factor of 2 to 3 would be necessary in most of the Central European countries, to accommodate increased fuel input prices and allow for necessary investment.

Such price increases will trigger energy-efficiency investment by end-users, such as industry or households. They may however exceed the financial possibilities of certain consumer groups. Governments will need to implement special programmes (e.g. direct subsidies) to cushion the price impact for those parts of the population which cannot afford to pay for their basic energy needs.

Pilot schemes in Poland showed that the refurbishment of the building shell and heat supply system, together with incentive tariffs, can reduce heat consumption by more than 20 per cent. Furthermore, the transfer of ownership of housing from the state to the private sector, which has commenced in several Central European countries, will generate an additional momentum for measures to reduce individual energy expenditures. Changes in ownership may eventually lead to improvements in energy efficiency as owners and/or occupiers can benefit directly from investment in thermal insulation and modernization of the heating system. However, the currently low level of investment in the building stock and the generally slow turnover of the capital stock in the residential sector will entail that such efficiency improvements are only slowly translated into demand reductions.

As in Western Europe, household energy demand has also increased because of market penetration by electric appliances, which has boomed in recent years. These appliances are typically smaller than their Western counterparts, but they are also less efficient. Furthermore, the ownership levels are below the levels observed in Western Europe. In Poland, for example, the combined ownership of refrigerators and freezers is approximately 1.2 per household, compared with 1.4–1.5 in Western Europe (IEA, 1995a).

Energy-efficiency improvements are likely to be substantial, particularly through measures to enhance the insulation of the building shell and the performance of the heating system. On the other hand, the residential sector of Central Europe is likely to see changes in the lifestyle similar to those in Western Europe, once disposable income increases: the demand for new electric appliances, the trend towards better equipped houses, increased level of comfort, and behavioural changes demanding more living space will offset the expected efficiency improvements.

2.2. Demand-Side Policies

Demand-side (energy-conservation) policy covers a wide range of measures which aim to improve technical efficiency in energy use and to reduce growth in energy demand. Energy authorities normally consider both energy pricing and non-price measures to be within the domain of energy-conservation policy.

Energy Prices and Taxes

Energy prices—besides economic activity and personal income—are the most important determinant for energy consumption. Prices undoubtedly induced most of the energy-efficiency improvement that took place in Europe after the first oil-price shock. Escalating energy prices during the 1970s triggered immediate energy conservation measures, such as changes in behaviour, and eventually further reduction in energy consumption through investment in more energy-efficient equipment or industrial restructuring. Prices result from international price movements, national price regulations, and taxation. Consequently, most of the literature does not regard pricing policy as an energy-conservation measure.

Energy taxes, which are sometimes three times the prices before taxes, for example for gasoline in the Nordic countries, have a decisive influence on consumers' choice of energy, on energy efficiency, and the environment. Taxation of energy has been driven only to a modest extent by energy policy considerations and is often pursued for budgetary reasons. Oil is taxed much higher than the other energy products, natural gas and coal, see Figs 2.9 and 2.10. On the other hand, most Western European countries do not tax electricity.

Average taxes on oil in the OECD have substantially increased since 1985 and current taxes on oil in the industrialized countries are now higher than the costs of imported oil. There has thus been a redistribution of the value of oil from producers to consuming countries. Though it can be argued that govern-

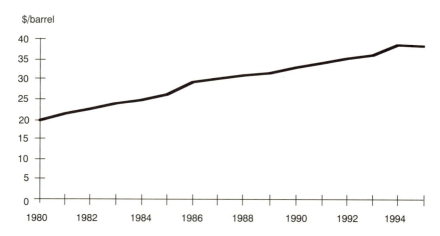

Fig. 2.9. Average taxes on oil in Western Europe

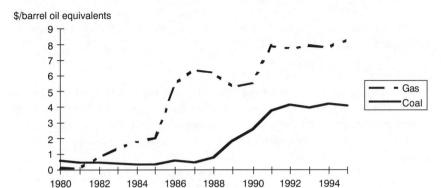

$/barrel oil equivalents

FIG. 2.10. Average taxes on gas and coal in Western Europe

ments wanted to maintain a high end-user price level for reasons of security of supply, it is more likely that budgetary considerations were the main motives. The highest tax levels are observed on gasoline, a category which is comparably inelastic in its response to price increases. The following section discusses non-price policies that governments have implemented since 1973, why these instruments have been implemented, and assesses, as far as possible, their impact on energy demand.

The Rationale for Energy-conservation Policy

Energy-conservation policy has for more than twenty years been considered an important component of energy policy. The rationale for an active policy to promote energy efficiency has shifted over time, from the concern for energy as an exhaustible resource which was predominant in the early 1970s, to the energy security issue after the oil-price increases in the 1970s. In the mid-1980s governments also began to pursue demand-side policies in addressing global environmental concerns.

However, the most powerful argument put forward in support of specific policy measures to conserve energy is the existence of the *efficiency gap*. This is defined as the amount of energy that can be supplied through energy-efficiency improvements, which have a higher rate of return than the cheapest alternative of supplies provided through expansion of energy production capacity. The proposition that an efficiency gap exists has scientific backing from a large number of engineering studies showing that energy consumption theoretically can be reduced by 20 per cent or more at a profit. Whether the efficiency gap actually can be harvested at a net social benefit is a subject of much debate, and it is primarily for this reason that the merits of energy conservation policy have been and still are controversial.

How Big is the Energy-efficiency Gap?

Most international estimates of the efficiency gap are in the range of 10 to 30 per cent of total energy consumption. Studies from the mid-1980s tend to show a higher potential than more recent estimates. Various studies surveyed in IEA's 1987 energy conservation policies study (IEA, 1987*a*) indicate that the potential in Western Europe, as of the mid-1980s, was around 20 per cent. A more recent study (IEA, 1991*b*) contains a less optimistic view of the overall size of the efficiency gap, see Table 2.1.

To some extent, the variation in estimates can be explained by differences in

Table 2.1. *Energy-efficiency potential: summary of opportunities and barriers*

	A Estimated share of total final consumption (%)	B Total savings possible[a] (%)	C Existing market/inst. barriers[b]	D Potential savings not likely to be achieved[c]
Residential space heating	11	10–50	Some/Many	Mixed
Residential refrigeration	1	30–50	Many	10–30
Residential lighting	1	over 50	Many	30–50
Industrial motors	5	0–30	Few/Some	0–10
Chemicals	8	0–10	Few/Some	0–10
Pulp and paper	3	0–10	Few/Some	0–10
Passenger cars	15	30–50	Many	10–30
Goods vehicles	10	30–50	Some	0–10

How to read this table: e.g. for lighting, over 50% per unit savings would result if the best available technology were used to replace the average lighting stock in use today over the next 'ten to twenty years'. Some of these savings would take place under existing market and policy conditions. But due to market and institutional barriers, there would remain a 30–50% potential for savings that would not be achieved.

[a] Based on a comparison of the average efficiency of existing capital stocks to the efficiency of the best available new technology. This estimate includes the savings likely to be achieved in response to current market forces and government policies as well as those potential savings (indicated in Column D) not likely to be achieved by current efforts.
[b] Extent of existing market and institutional barriers to efficiency investments.
[c] Potential savings (reductions per unit) not likely to be achieved in response to current market forces and government policies (part of total indicated in Column B).
Source: IEA (1991*b*).

methodology and assumptions about future energy prices, and the lifetime of the equipment. The estimates shown in Table 2.1 are based on lower price trajectories than the studies referred to in IEA's conservation study from 1987. This can explain, at least in part, the downgrading of the estimates. Furthermore, earlier estimates were possibly too optimistic and overambitious in their assessment of a cost-effective efficiency potential which could be realized without substantial government intervention. In the second half of the 1980s governments were more and more inclined to pursue energy conservation with less vigour than in the past.

As in earlier studies the potential is larger in the residential sector than in industry. However, the IEA study indicates that there may not be any potential for efficiency gains in industry that go beyond the improvements that will be achieved by market forces. A lower energy-efficiency gap in industry is explained by the fact that industry generally has the incentive to reduce its production costs. Industry would therefore undertake energy-efficiency investment to the extent that this was cost-effective. The potential in residential energy use and in private transportation ('passenger cars') is considered to be significant (10–30 per cent). Still it is important to keep in mind that purchase of cars and household equipment and appliances are not only governed by economic considerations but also by personal taste and perceptions of comfort. 'Suboptimal' investments from a purely energy-efficiency point of view can, therefore, in these cases make good economic sense for the consumers. The different size of the efficiency gap in different end-use sectors is explained by the existence of market imperfections or barriers to energy efficiency, which are generally larger for households and in the transport sector than for industry.

Whatever methods and assumptions are applied, a considerable number of investment projects exist that have a high rate of return, but which are not pursued by energy consumers. This has forcefully been put forward as an argument for a revitalization of energy-conservation and demand-side policies. Some analysts, however, dispute the existence of an energy-efficiency gap altogether on the grounds that most of the barriers to energy efficiency as discussed below do not exist. Such barriers, they argue, merely reflect unaccounted (transaction) costs or simply result from the consumers' liberty to choose freely his/her convenience and service levels and willingness to accept a higher energy bill for their personal taste or lifestyle (see *Energy Policy*, 22/10). The energy-efficiency gap is often used as an argument to pursue heavy-handed energy conservation policies, such as standards and regulation.

Energy-conservation Measures

The justification for energy-conservation policies is that they are needed to narrow or eliminate the efficiency gap caused by market barriers and imperfections. The reasons for the efficiency gap have been described extensively in the literature, see for example IEA (1991*b*, and 1994*c*), Jochem and Gruber (1990), and Grubb (1990). Three sets of factors are noted as significant:

- Lack of information and technical skills lead to investments and operational dispositions that are economically suboptimal and result in disproportionately high energy use.
- Consumers' required rates of return on energy-saving investments can be higher than comparable discount rates applied for energy-supply investments. According to some surveys of low-income households investments are required to pay back in less than one year in order to be attractive.
- Separation of costs and benefits: for example, tenants may be discouraged from renovating buildings if only the owner receives the benefit. Lack of cost-effectiveness in public-sector investments and pure operational performance in public buildings can also be included in this category.

In energy conservation policy four categories of measures have been of particular importance: information and exhortation, regulations and efficiency standards, financial incentives, and demand-side management by electric utilities.

Information and Exhortation

Information campaigns have been primarily addressed to private residential consumers and the transport sector, and consumer associations have often been involved. Energy utilities also play a central role in disseminating information on energy efficiency. In most countries utilities are mandated by law to provide consumers with information and advice on the most cost-efficient energy-supply options, including options for investment in energy conservation. The aim of information campaigns has not only been to promote profitable energy-efficiency improvements, but also to encourage consumers to save energy, i.e. reduce the consumption of energy-using services and to change their behaviour. For example, general campaigns that appeal to 'green consciousness' have aimed at lowering demand for transport services and energy services in households.

Labelling of appliances tends to be less costly than information campaigns, and is mostly undertaken on a voluntary basis. Equipment vendors sometimes provide information on the expected energy costs, potential savings, or efficiency (e.g. data on car fuel efficiency). In Europe only Sweden has introduced a comprehensive mandatory labelling system for electric appliances. However, the European Union has recently enacted a directive on energy labelling of electric ovens and others may come as part of the SAVE-programme (see Chapter 8). Energy auditing schemes and the supply of other technical information to industrial managers have had good results in many European countries.

The impact of energy labelling on individual purchase behaviour is thought not to be significant (see Schipper, 1987, IEA, 1991*b*). One reason for this is that energy costs rank very low as a criterion to be considered in decision-making. For example, although most car manufacturers still provide information on fuel efficiency, consumers tend to be more interested in the car's comfort, colour, and power. Although information campaigns have been considered a cornerstone in energy conservation policies in Western Europe the results appear not to have been significant. Broad general campaigns have tended to give meagre results whereas campaigns that are targeted at specific consumers and information that is concrete in prescribing change in behaviour have shown to be more effective (see IEA, 1989*a* and 1991*b*).

Focused information and training for smaller groups are labour-intensive and costly. Still, evaluation of such programmes shows that they can represent a cost-effective way to promote energy efficiency. The best prospects for successful information programmes are within industry and service sectors if such programmes are carefully targeted, despite relatively small efficiency gaps in these sectors. In addition, since small-scale programmes are more efficient than larger schemes, the aggregated effect of information on energy demand will probably remain modest.

Regulations and Efficiency Standards

Voluntary and mandatory technical standards are applied to improve energy efficiency for specific transport vehicles, machinery, appliances, or buildings. Standards are meant to compensate for energy consumers' lack of information or other market barriers, such as the separation of costs and benefits. Due to the large efficiency gap in household and service sectors, building codes are the most important set of standards to have been enforced in Western Europe. Some 20 per cent of total energy demand can be addressed by such measures. New buildings today typically use half the energy needed in the early 1970s, and this development is to a large extent the result of building codes (IEA, 1991*b*). Some countries, like Denmark, Germany, Switzerland, or the United Kingdom, have begun in the 1990s to develop new building codes mandating higher insulation levels, or to implement performance standards that require a certain level of energy performance, but leave the design details to the architect.

There has been no mandatory fuel efficiency standard for road vehicles in Europe like the corporate average fuel economy (CAFE) standard in the USA. Some countries have mandatory car inspections and speed limits primarily for safety and environmental reasons, and these may also have positive effects on fuel efficiency and energy demand. Energy authorities in car-producing countries, such as Germany and Sweden, have agreed voluntary targets for efficiency improvements with national authorities, and these targets have been more than fulfilled. But it is difficult to assess the specific effect of these voluntary agreements. There are indications that the efficiency improvements stipulated would have been achieved by the market in any case (see IEA, 1991*b*). In fact, in 1978 the German car manufacturers agreed with the government to reduce fuel consumption by 15 per cent by 1995, and this target was exceeded by 8 per cent.

Appliance and equipment standards, for example for residential appliances, were first introduced in the USA, but are less common in Europe. In 1991 the European Commission began to review the possibility of EU-wide standards. A directive under preparation is expected to establish maximum consumption levels for freezers and refrigerators introduced in 1997, some 10 per cent below the 1991 market average. The Commission also plans to develop a directive on labelling dishwashers, washing machines, and dryers, after which it will consider the possibility of standards for these appliances. Though such an approach is favoured by some countries in the EU, such as the Netherlands and Denmark, Germany for example has so far been reluctant to accept the introduction of appliance standards. Arguments used against the standards are that they burden unduly the industry whose financial viability is more and more squeezed by competition from outside the EU, that they constitute a market interference, or that they require a detailed technological and market knowledge which is usually not at the disposal of the regulator.

Financial Incentives

Financial incentives are targeted towards energy consumers who abstain from efficiency investments due to high payback requirements or budget constraints. This category covers a large number of instruments such as grants, interest subsidies on loans, and tax incentives. They are often applied in combination with information programmes. Some of the demand-side management (DSM) activities by energy utilities include financial incentives. These are discussed below.

Financial incentives were widely used at the end of the 1970s and in the first part of the 1980s. By the mid-1980s some of the larger programmes, in Sweden, Denmark, and Germany, were scaled down, partly for budgetary reasons and partly because the programmes were less successful than had been anticipated. Declining energy prices further reduced consumers' and governments' interests in the programmes.

However, in the early years of the 1990s financial incentives have had a renaissance as a tool in environmental policies. Some countries continue to undertake financial incentive programmes on a relatively modest scale. Others, such as Austria, the Netherlands, Italy, or the United Kingdom, still apply significantly sized programmes, including tax deduction for building-shell improvements, grants for renewable energies or co-generation of heat and electricity by industry, or subsidies for manufacturers to produce highly efficient appliances, such as refrigerators. But financial incentives provided by governments have changed in recent years and represent a policy shift from trying to create a market to motivating market participants to make a better choice. First, some grant programmes have been discontinued, or streamlined and targeted to achieve, for example, environmental objectives. Second, there has been a trend towards using limited public resources to encourage private capital, rather than providing complete public financing (see IEA, 1994*e*).

A new approach has been third-party financing, in which an energy service company knowledgeable about technology and processes provides financing, equipment, and expertise to reduce energy use in buildings or industrial plants. In return, the energy service company is paid by the owner of the building or plant for a share of the energy cost savings. Similar arrangements which are based on commercial and contractual relations between end-users and know-how providers, and which are often undertaken in conjunction with financial institutions, include equipment leasing and energy performance contracting. These approaches for energy-efficiency contracting have the advantage that they are based on the principles of the market, allow for adequate sharing of risk among participants, and eliminate the problem of 'free riders'.

Experience has shown that financial incentive programmes can be very expensive if not properly targeted. In some cases, the free-rider effect (recipients of financial support who would have acted even in absence of the support) resulted in a misallocation of scarce public resources. In other cases, such programmes have led to unnecessary transfers of payment and cross-subsidies (e.g. some utility DSM programmes). A Norwegian programme was discontinued in 1994 after an evaluation showed that grants offered attracted a large number of 'free riders' (see Energidata, 1993 and Haugland, 1994). In the Netherlands, subsidies for combined heat and power plants are being phased out as they contributed to the construction of decentralized generating capacity, leading to the build-up of overcapacity and instability in the public-supply system.

Demand-side Management by Utilities

Since the mid-1980s, electric utilities in some Western European countries have begun to promote to their customers end-use efficiency. Such activities require a departure from the traditional role of utilities as supplier of electricity. The development of DSM was in part a response to rising supply costs and demand uncertainty. It was also partly due to the pressure on utilities to promote energy efficiency as one option to lower demand growth or to reduce the environmental impact of electricity production.

DSM activities include technical advice, information campaigns, energy audits, give-away or rebate programmes for new and more efficient technology, such as high-efficiency light bulbs or refrigerators. In the residential sector, automated regulation of space heating, insulation of water systems, interruptible appliances, and lighting programmes have been tried. In the service sector, building-shell retrofits, improvements in space conditioning equipment, and the installation of efficient lighting systems have been implemented. In the industrial sector, utilities have offered special rates to increase the incentive for industrial customers to reduce use of electricity during peak-load hours.

Utilities in many European countries implemented various initiatives involving subsidies for the installation of energy-efficient technology, such as rebate or give-away programmes for light bulbs or refrigerators. These programmes were mostly one-off initiatives and not done on a continuous and systematic manner. If not carefully designed, targeted, implemented, and monitored, the free-rider effect in such programmes can be substantial, leading to higher electricity costs for non-participants. There are doubts whether utilities operating in a competitive and deregulated environment, such as in the United Kingdom and Norway, will pursue DSM. The incentives for utilities to undertake DSM may disappear in an environment where electricity prices are set competitively. Governments in some countries require utilities to prove that the investment they propose will provide electricity at the lowest possible cost, also taking account of demand side measures. This approach is often called integrated resource planning.

The Effects of Energy-conservation Policy

Indisputable estimates of the impact of energy conservation policies on energy demand do not exist. Nor is a comprehensive quantification available of the net economic benefit of such instruments, for cases where public funds were used. The benefits would have to be assessed against the economic losses that result from tax increases which would be required to finance certain energy conservation policies. Economic analysis suggests that raising public funds through taxation creates economic distortions that could amount to 50 per cent or more of the fund spent (see Ballard and Fullerton, 1992).

Though there is no doubt that certain energy-efficiency measures, such as building codes and targeted energy-information campaigns have achieved some results, their overall impact on energy demand has been limited. Building codes address a significant portion of energy end-use, but they cannot achieve quick results, given the slow turnover of the building stock. Furthermore, the enforcement of more stringent codes for existing buildings is costly. In particular, for the fastest growing end-use sector, transport, energy policies other than prices and taxation have shown little results.

Targeted information and awareness programmes, with financial incentives, had some success. However, the latter has also led to significant costs and unwanted side-effects, such as free-riders. Given budgetary constraints, it is unlikely that governments will in the future have recourse to these programmes on a large scale. New schemes such as energy-efficiency contracts, based on principles of the market, can help exploit the cost-effective efficiency potential in some sectors, such as buildings and industry.

3
Energy-Sector Developments: Oil

The oil industry, more than other energy sectors, is global in its character and operations. The geographical concentration of reserves and the vital role of oil in modern society has made it the principal commodity in international trade. Oil is, however, more than an internationally traded commodity; it is a strategic resource. The political importance of oil has a long history and encompasses numerous political and military conflicts. Even though the oil market has evolved from being dominated by a small number of vertically integrated companies to being highly competitive, the political dimension of oil continues to have a major bearing on energy policy.

The forces released in response to the oil-price shocks of the 1970s demonstrate the impact of turbulence in the international oil market on energy markets and policies. The oil-supply disruptions led to three types of responses. First, there were the energy supply-and-demand responses, mainly through price-induced market reactions, but also by direct policy measures promoting fuel substitution and energy conservation. These issues are discussed in Chapters 6 and 2. Secondly, there were internationally coordinated initiatives to alleviate the consequences of external disruption in oil supplies. This is discussed in Chapter 7. Finally, the oil shocks had major implications for the whole structure of the oil market, both internationally and at the national level, and on how governments interacted with the oil industry, which is the theme of this chapter.

3.1. Historical Background

Already early in the twentieth century, access to oil reserves was an important part of European colonial policy. In 1914 the British Government secured access to oil reserves through the purchase of the Anglo-Persian Oil Company, later known as British Petroleum (BP). In 1924, Compagnie Française des Petroles (CFP-Total) was formed as a state-owned French company. Together with the large US companies, Exxon, Mobil, Texaco, Chevron, and Gulf, and the Anglo-Dutch Royal Dutch Shell, these eight companies became the main international oil players (majors). Royal Dutch Shell, the third and the biggest of the European based majors, was used less than the other companies as an instrument for strategic and political interests.

Until the second part of the 1950s the majors controlled practically all the oil supplied to Europe. In 1950 their share of world oil trade was 98 per cent, from where it gradually declined to 89 per cent in 1957 and 78 per cent in 1966 (Percebois, 1989). Despite the gradual development of alternative trade channels, the bulk of international oil trade operations continued to take place within the subsidiaries of the oil majors, often at prices adjusted internally to minimize tax exposure. By 1972 approximately 8 million barrels a day (mbd) out of the 25 mbd traded by the majors were sold to other oil companies, primarily European, at full price—i.e. the posted price in the Middle East plus transportation costs (Cowhey, 1985). The availability of crude oil from the Middle East, the comparative cost advantages of the oil multinationals and the expansion of European economies explain the majors' increasing involvement in the European refinery sector between 1960 and 1973. For example, Europe's share in total US oil company investments increased from 12 per cent in 1950 to 28 per cent in 1970 and 46 per cent in 1973.

Europe's mounting energy deficit, driven by fast growth in oil imports and decline of domestic coal production and demand, caused some political concern. However, the increased dependence on imported oil was not counteracted by policy measures since the dominating political view was that Europe's role in the international division of labour should be to reinforce its position as importer of raw materials and exporter of manufactured products. Furthermore, the US government gave assurances to other OECD countries that in case of supply contingencies, its oil production could be increased and its stockpiles be used as back-up (Shaffer, 1983). Europe's crude oil imports, mainly from the Middle East and Africa, were thus allowed to grow.

In the early 1950s heavy fuel oil was the dominant product in European oil demand. It continued to grow rapidly through the 1950s and 1960s. In the 1970s the increase took place particularly in the electricity sector, where by 1973 oil accounted for more than one-third of fuel use. During the 1960s, however, light products had an even stronger growth. From 1960 to 1973, gasoline consumption increased by 200 per cent and that of gas/diesel oil by 330 per cent. By the early 1970s Europe had become critically dependent on oil: 60 per cent of energy consumption was oil, virtually all of which was imported, refined, and distributed by the majors.

The oligopoly power of the majors in world markets started to deteriorate in the mid-1960s. The resulting disruptions in market structures gradually led to sharp competition between the majors. For example, while the US majors expanded operations in Europe, the European majors, with vast production in the Middle East, tried to increase sales in US markets. The faltering cohesion between the majors deteriorated further as they no longer felt convinced about the benefits of being barred from each other's 'exclusive' oil-producing countries. Finally, limited capacity to expand integrated operations in pace with demand paved the way for national and independent oil companies to increase their market shares, both upstream and downstream. By 1970 about 20 per cent of oil output in exporting countries was produced by about 200 American, European, and Japanese companies other than the majors. All these factors convinced governments in oil-exporting countries that the time was ripe to demand a larger share of the economic benefits from petroleum extraction.

3.2. Prices, Demand, and Supply

Crude Oil Prices

The growing national self-confidence in the Middle East, the erosion of the majors' dominant position, and the continuous high growth in world oil demand led to an upward pressure on crude oil prices. Libya and Iran demanded, and got, higher crude oil prices from the international oil companies. From January 1970 to September 1973 (prior to the Yom Kippur War) prices for Saudi Arabian Light increased from 1.40 $/barrel to 2.70 $/barrel. When the war broke out in October 1973 and the Arab arm of OPEC imposed a selective oil embargo on Western countries, the first oil crisis was a reality and prices rose to 10.50 $/barrel by March 1974. Even if the price hike was more a consequence of a tight market than the result of political objectives, the impact on Western governments and public was profound.

The second oil-price shock in 1978–9 led to an increase in nominal crude oil import prices to Western Europe which peaked at 36 US$/barrel in 1981, thirteen times the nominal price level in early 1973 and 2.5 times price quotations prior to the second oil-price shock in 1979. The slide in crude prices from 1981 and the subsequent price collapse in 1986 brought the price level down to 10 US$/barrel in July 1986 and to 15 US$/barrel for the whole of that year.

Oil-Product Prices

Oil-product prices underwent parallel, but less extreme, oscillations. In real terms average oil-product prices in Western Europe rose by 50 per cent from 1978 to 1982, whereas real crude prices doubled (see Fig. 3.1). Oil-product

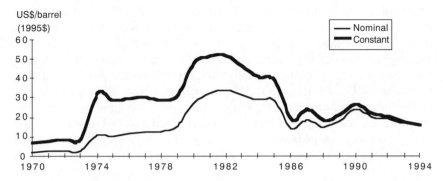

FIG. 3.1. International crude oil prices (nominal and constant US$/barrel)

prices rose less than crude oil prices because other cost components in oil products—such as refining costs, distribution costs, and taxes—rose less or not at all. The crude oil costs typically range from 15 to 80 per cent of the final end-use prices. For gasoline, for example, crude oil normally accounts for 20 per cent or less of the total cost. During the period of falling crude oil prices from 1982 to 1986 oil-product prices fell by only 25 per cent. The modest reduction was primarily a result of higher taxes on oil products. In particular, gasoline and light fuel oils were the target of increased taxation.

The rebound in crude oil prices after 1986 was the result of a revaluation of US$ versus most European currencies. This was not accompanied by any significant increases in oil product prices and thus indicates that the significance of crude oil prices on product prices had decreased. This in turn was the result of a higher tax component in the end-use prices of oil products, notably gasoline, and the larger weight of gasoline and other 'high value' products in the refinery mix.

Policy Responses and Market Reactions

The changes in oil consumption that followed the oil-price shocks were caused by price-induced market responses and macro-economic upheavals, and by policy action, including primarily changes in the fuel choices away from oil in electricity generation and oil-product taxation. Energy conservation policies (other than pricing policies) were also stepped up after 1973 but contributed much less to reduced oil use (see Chapter 2).

Policies to cut down dependence on oil imports were particularly effective in the case of fuel-oil use in the electricity sector. The tight regulation of this sector made it possible for governments to mandate fuel choices and investments which reduced oil's share (see Chapter 6). However, the most important part of the decline in European oil consumption took place in the industrial sector. Here the decline was, as described in Chapter 2, a result of structural changes and energy-efficiency improvements triggered by price increases. In the household/commercial sectors oil was partly substituted by natural gas or electricity. On balance, all of the decline took place in stationary uses while oil consumption in the transportation sector continued to rise. Total European oil consumption reached a peak of 15 mbd in 1973 and then a second top of 14.7 mbd in 1979, before it declined to a low of 11.9 mbd in 1985. Despite the fact that oil consumption has increased annually since 1986, these two peaks still remain historical highs.

The high level of oil prices and the supply diversification policies of importing countries also had an important spin-off effect on the development of oil fields in the North Sea, previously regarded by the oil industry as a high-cost area. The combined oil production of the United Kingdom and Norway increased from virtually nil in 1973 to 3 mbd a decade later and to as much as 5.4 mbd in 1995. These production levels had a significant impact on international markets as a growing share of the volumes previously imported by Western Europe from OPEC countries now had to be sold in other world regions. Furthermore, although most of the oil production from the North Sea was sold under long-term contracts between producers and refiners, the surplus was sufficiently high to provide the base load liquidity required for the building of a lively European spot market (Mabro, 1986, and Horsnell and Mabro, 1993).

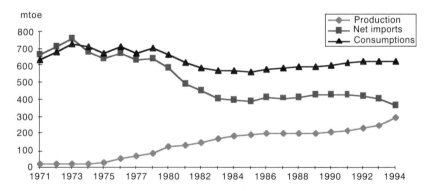

FIG. 3.2. Production, consumption, and net imports of oil in Europe

3.3. Industry Structures

Fragmentation of the Market

The turbulence of the 1970s and early 1980s led to a breakdown of the vertical integration of the oil industry. The majors' control of the entire vertical chain of oil supplies was broken up and a number of new enterprises entered the oil market. As the main producing countries in the Middle East assumed ownership of their reserves, the majors' own production of crude oil went from 25 mbd in 1972 to about 8 mbd in the latter part of the 1980s and they had to start purchasing oil from a variety of alternative sources. However, until the Iranian revolution in 1979, a large part of the crude supplies was still purchased on long-term contracts between the majors and the large producing countries. The deverticalization increased after the second oil-price shock. BP lost 40 per cent of its crude after the revolution in Iran and the nationalization in Nigeria. The company sought new crude supplies in the spot market. The downward pressure on oil prices from 1981, and reluctance of oil producers to adjust prices quickly enough, gave the spot market a central position in international oil trade. Hence within a time span of ten to fifteen years the integrated nature of crude oil supplies was broken. The majors had to reorganize and seek a new role within a different commercial environment which has all the main characteristics and instruments of a competitive international commodity market.

Structural Changes in Oil-product Markets

After the first oil-price shock and corresponding adjustments in oil demand during the remainder of the 1970s, Europe entered the 1980s with an excessive and largely anachronistic oil refinery sector. The most important deficiency was the lack of capacity to convert heavy fuel oil to higher value products. Prior to 1973 European refineries had been designed to yield a large proportion of low-quality fuel oil in competition with coal. After 1973 heavy fuel oil lost in competition with other energy carriers, primarily because of the high cost of crude oil but also to some extent due to taxes and stricter environmental regulations. On the other hand, light products, and in particular gasoline, increased their market shares despite large tax increases. Moreover, these market segments remained profitable, partly due to a deficit in European refineries to cover market requirements.

Already during the 1973–9 period European refineries were strongly affected by the decline in demand for heavy fuel oil and low-grade middle distillates. However, no significant measures were taken by oil companies,

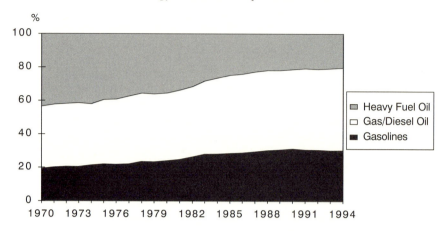

Fig. 3.3. Consumption of gasoline, gas/diesel oil, and heavy fuel oil in Western Europe

neither the state-owned European companies or the majors, to reduce their surplus in refining capacity. As a consequence, only 60 per cent of the distillation capacity in Western Europe was utilized in 1979. The state-owned European companies actually increased their operations and attained by 1986 a 32 per cent share of oil refining, up from 17 per cent in 1976. Their 'staying power' during a period of decline and stagnation in oil product demand is partly explained by government support, which made them less vulnerable to the consequences of commercial losses than privately owned companies.

When the disinvestment belatedly started in 1982 the majors led the way. The profits in downstream activities had dropped dramatically from 1980 onwards. On a global basis, downstream profit of the majors dropped from $9.6 bn in 1980 to $1.7 bn in 1984 (IEA, 1987*b*) and the development was probably more serious in Europe since the refinery margins at the time were smaller than in other regions. The delay in capacity reduction was due to the

Table 3.1. *Refinery ownership by type of company in Western Europe (% shares)*

	1976	1980	1986
Majors	61	48	40
State-owned European	17	28	32
State-owned non-European	—	—	3
Independents	22	24	25
TOTAL	100	100	100

Source: IEA (1987*b*) and Bacon *et al.* (1990).

considerable resources available to the large companies, enabling them to hold out for a long time and their continued, if diminished, vertical nature. They saw refining as an important part of a vertically integrated structure, and were willing to subsidize the downstream operations with profits made elsewhere. However, the erosion of downstream profit became significant and the value of vertical integration was less evident.

Structural Change in the Organization of the European Oil Industry

The overall structural change that has taken place during the last two decades can be summarized as follows.

1. A substantial decline in oil consumption has occurred in stationary applications. The bulk of oil consumption increase in Europe is in the transport sector where oil enjoys a virtual monopoly and where no fuel alternatives are currently commercially available.

2. Net import of oil to Europe was reduced from 14 mbd in 1978 to 8 mbd in 1994. The value of oil import has also declined substantially; in the European Community from 26 per cent of total imports in 1980 to less than 10 per cent in 1992 (Mitchell, 1994).

3. The European refinery sector has been radically restructured to match demand trends and the share of European oil companies' involvement in European oil markets has increased substantially. The process of modernization in the refinery sector has evolved faster in northern Europe than in southern Europe.

4. All European countries have deregulated oil import prices and domestic retail prices. Prices are basically determined at the world market. Moreover, the impact of the majors on oil-market developments is significantly reduced by the emergence of national and foreign oil companies, both state-owned and independents.

Although these trends have affected almost all European countries, policies towards the oil sector have shown some notable differences between countries. This is reviewed below for some of the major European countries.

3.4. National Trends and Policies

Germany

Until the mid-1960s Germany protected its own domestic oil production through high import tariffs. The policy was abandoned when a common structure of external tariffs was established within the European Community. As a result foreign companies (e.g. Shell, BP, Exxon) became the dominant suppliers of imported crude oil and also took over most of the remaining production of oil in Germany. However, before the oil shock in 1973, concern for security of supply led to the formation of a German exploration company, Deminex (Deutsche Erdolversorgungsgesellschaft).

Initially Deminex was owned by eight large enterprises with interests in the energy sector, but eventually VEBA, the largest of them, gained control over the company through active support by the Federal Government and despite persistent opposition from the Cartel Office. From 1969 to 1986 Deminex received DM 2.6 bn in subsidies (Lucas, 1985). However, production of crude oil in Germany declined from 0.13 mbd in 1973 to 0.08 mbd in 1983 and the company failed to make any profit on its upstream activities until 1982. These very low levels of production can explain why the German oil industry has been organized as an open market with few restrictions for foreign operators, and why the development of German oil companies has primarily been linked to refinery operations.

The downstream part of the German oil sector has for all practical purposes remained a free market through the two oil crises until now. There have been no price controls and a large number of companies have had operations in refining and distribution. The refining capacity was reduced in line with lower oil demand. In the ten years from the end of the 1970s refinery capacity was reduced by 50 per cent, far more than in France, the United Kingdom, and Italy. Over the same period the majors' share of the capacity fell from 61 to 55 per cent (see Schiffer, 1988). It is symptomatic of the open competition in the German downstream market that the country has had substantial net imports of petroleum products through the entire period since 1973, and there has been no policy to shelter national refineries from competition.

France

France is poorly endowed with oil and has an even more modest indigenous production than Germany. In 1994 the domestic oil production of 80,000 bd covered only 3 per cent of national consumption. France is thus highly dependent on oil imports, which explains why successive governments have given priority to nuclear electricity as a domestic source of energy. Still, France is the home of two large oil companies, CFP-Total (TOTAL) established in 1924 and Elf-ERAP (Elf). Both companies spread a major part of their exploration and production activities across many countries, though their refinery operations are highly concentrated in France. Since it was formed, TOTAL has operated in Iran and Iraq, along with the other majors. Elf initially had major oil concessions in Algeria but the company's engagement was substantially reduced after a dispute over royalty payments and nationalization of oil in the late 1960s. To compensate for loss of crude reserves in the Middle East and the Sahara both French companies sought crude oil in other regions, for example in the North Sea, West Africa, South-West Asia, and North America, but France soon had to give up its objective of having control over crude oil production equivalent to its domestic oil use.

Another objective in French energy policy was to keep 50 per cent of the domestic oil-product·market for French companies. This target was also abandoned in the course of the 1970s. In 1978 the French Government started to liberalize petroleum product prices and import quotas were removed in 1979. Price controls on diesel and petrol remained in place until 1985, contributing to the record losses in refining in 1981 and 1982. Of a total loss of FFr 13 bn in the refinery sector in 1981, some FFr 8 bn were accounted for by Elf and TOTAL (Lucas, 1985). This situation accelerated the legislative process to replace the 1928 oil law with a new regime, aimed at reducing the government's ownership of oil companies while maintaining control of supplies should a crisis necessitate government intervention. The new law was passed in 1993. Besides removing the requirement for government authorization to import crude or oil products, the Act also provides regulations to safeguard the availability of oil supplies. Examples of this are that refiners must have access to French-flag tankers equal to 8 per cent of their refinery throughput; the government has the right to veto all refinery projects; and security stocks required to be held by companies will be controlled by a government-led committee. The new law also provides for the government to introduce measures forcing refineries to cultivate raw materials for biofuels.

Emerging since 1979, a curious development has been the rapid growth of independent petrol distribution chains, originally owned by wholesalers of foodstuffs and thus called 'supermarkets'. Profitability of these distribution chains is based on high throughput, low operational costs, and large price discounts to attract customers to the shopping centres (malls) often located on the outskirts of large cities. Their petrol is imported. When the 'supermarkets' appeared, TOTAL and Elf calculated that their refineries gave them an advantage over independent importers because, until 1981, France was still a net exporter of refined products. However as trade restrictions were gradually removed, the dominant position of Elf and TOTAL began to recede. By 1989 the share of 'supermarkets' had reached 38 per cent of the distribution market. Refineries reacted by increasing their production of unleaded petrol, introduced to France in 1989, which was not available at the supermarkets. This example underlines the difference between oil and other energy markets, and the similarity of the former to other, non-energy commodities. It is hard to envisage supermarkets entering into retail distribution of coal, gas, or electricity.

Today the government considers that the size of the two French oil companies is satisfactory, nationally and internationally, and sufficiently large to survive without state support (French Ministry of Industry and Foreign Trade, 1993). TOTAL was privatized in 1992 and the state shareholdings in Elf were sold in 1995–6, though the state has retained a 'golden share' to secure that the company develops strategies in accordance with French interests. In TOTAL the state retained 15 per cent.

United Kingdom

The British Government was among the first to underline the strategic importance of oil when it purchased the Anglo-Persian oil company, later known as British Petroleum (BP). BP grew to become one of the majors, with fully integrated exploration, transportation, refining, and distribution activities. BP did not at any point in time hold a monopoly in UK oil sector nor did the government use BP as an active instrument in domestic energy policy, though it promoted the company to play a key role in the development of oil fields in the North Sea. Oil was discovered in the UK continental shelf in 1966 and the first commercial oil find was made three years later. By 1978 UK production reached 1mbd, then doubled in 1982 before finally reaching a peak of 2.64 mbd in 1985. It subsequently fell to less than 2 mbd by 1991, but has since grown markedly and was by 1995 at 2.6 mbd.

Following the campaign of the Thatcher Government to reduce state ownership in industry, virtually all state holdings in BP were sold in 1987. The measure was based on the Oil and Gas (Enterprise) Act of June 1982, which paved the way for the privatization of state assets. It is worth noting that in 1987, when the Kuwait Investment Office (KIO) decided to buy 21.7 per cent of BP's shares, the British Government intervened and managed to force KIO to reduce its holdings to 9.9 per cent, in order to avoid single investors controlling the company.

Another national oil company, central to the development of the shelf, was the British National Oil Corporation (BNOC), established in 1975 to secure the state's interest in North Sea production. This company was in fact entitled to buy 51 per cent of British production. BNOC had an important price-setting role in the international oil market because it announced the price at which they would buy oil from operators and set the price for crude offered to refiners. During the second oil crisis in 1979 BNOC was actually at the forefront in raising crude oil prices, while seven years later in 1986, as oil prices collapsed, the company lost money as it had purchased oil from North Sea operators at a higher price than the subsequent selling price. This contributed to BNOC being abolished (Yergin, 1991). The company was no longer perceived as having any effective function in supporting the public economic interests in relation to oil production in the North Sea and there were not considered to be any energy security arguments for maintaining it.

After the sale of BNOC assets the UK became one of the least demanding oil-producing countries in licensing terms (Ross, 1987). Before the abolition of BNOC the upstream market structure in the UK was dominated by six equal-sized firms. After 1985 this number gradually increased to eleven. After this, all majors were soon represented on the shelf, as well as numerous US oil independents such as Conoco, Occidental, and Amoco, and European companies such as Elf, Fina, Norsk Hydro, Statoil, Veba, Neste Oy, and AGIP, to name a few. The rapid increase in oil production during the 1980s made the UK an important exporter of crude oil, most of which went to the USA and the Continent.

Today the UK is a country with a highly developed and well-diversified upstream oil industry. Seventy-four oil and gas companies are listed on the London stock exchange. Annual investments of US$7.2bn in the oil sector, fully undertaken by private companies, represented 20 per cent of total domestic industrial investments in 1993. In 1992 upstream taxes and royalties generated US$2.2bn, in addition to North Sea company payments of US$1bn in corporation tax (Mitchell, 1994). Furthermore, following a tax reform package for the upstream sector passed during 1993, there has been a renaissance in oil output, bringing production in 1995 close to its peak level of 1985.

Oil refining in the UK has been strongly linked to competition with coal in the electricity sector and with gas and coal in all other stationary applications. For example, until recently UK legislation allowed use of fuel oil with a maximum sulphur content of 4 per cent. This is a very high ceiling, primarily established for regulating coal, and easily met by UK refiners since oil from the North Sea only contains 2.2 per cent sulphur. Even power plants based on imported heavy fuel oil have rarely consumed hydrocarbons with sulphur levels above 3.5 per cent. Sulphur and nitrogen oxide target reductions, adopted under the EC Large Combustion Plant Directive, were less severe for the UK than for other EC countries in recognition of the significance the coal industry had in the UK economy.

Italy

Italy did not succeed in securing the same access to oil in the first part of this century as the United Kingdom did with BP and France with TOTAL. The state-owned company AGIP (Azienda General Italiana Petroli) was formed in 1926 and is now an incorporated company to the state industrial conglomerate Ente Nazionale Idrocarburi (ENI). AGIP's share of the domestic market remained at about 25 per cent from the Second World War until 1973 (Lucas, 1985). Today AGIP's exploration and production activities are spread across twenty-four countries, the most important being Algeria, Angola, China, Egypt, Nigeria, Norway, and the UK. Domestic oil production has stayed at around 90,000 bd since 1988. Most of Italy's upstream activities are carried out by AGIP. Production is strongly centred in the Po Valley and the Adriatic Sea, though gradually moving towards the south. As with other European oil companies, AGIP's refinery activities are concentrated in its home country.

In 1973 Shell and BP left Italy, partly due to restrictive price control by the Interministerial Committee for Prices (see Lucas, 1985). ENI was forced to take a larger market share to compensate for the loss of Shell and BP supplies. Price control was temporarily relaxed after the second oil-price shock, as it became increasingly difficult to find oil for the Italian market. However, delays in adjustments of controlled petroleum product prices led to major losses in 1981 and made Amoco seek buyers for its operations in Italy. Eventually price control had to be eased.

Decades of rigid price controls left the Italian refinery industry in a poor position to finance investments for the processing of high-quality products. The country's state-owned refineries are now being guided towards privatization while price controls are being removed. Oil companies from Kuwait and Libya, which together supply almost half of Italy's crude oil imports, have been among the most active purchasers of downstream assets. So far the Libyan oil company Tamoil owns a refinery at Cremona and 2,300 filling stations; Kuwait Petroleum owns a simple refinery at Naples which it uses as import terminal.

Oil consumption of nearly 2 mbd in 1995 represented 60 per cent of the Italian energy market. The possibilities of reducing this share are limited due to the country's decisions not to build more nuclear plants and to reduce coal consumption. The remaining fuel alternative is gas, the share of which is being increased with imports from Algeria, Russia, and the Netherlands. As the cost of transporting gas from distant sources is high, the Italian refinery industry has not been exposed to strong competition from this fuel.

The large size of the Italian refinery industry has roots in the 1960s and

1970s. Government policy was to render Italy a major centre for crude oil processing, by importing crude from Africa and the Middle East and exporting oil products not needed in Italy to other European countries. Refinery capacity, mainly located in the Mediterranean, reached a peak of 4.13 mbd in 1980, falling to 2.24 mbd in 1993. The problem at present is that the country still relies upon a sizeable share of small refineries which are just about the economic minimum to justify investment in upgrading facilities. On the other hand, the largest refineries are quite efficient as they yield a high proportion of light and middle distillates from a large variety of crudes. However, despite the large size of the domestic refinery industry, Italy imports light distillates as well as fuel oil for power generation. The balance of oil-product imports and exports changes from year to year depending on total European supply and demand.

The challenge ahead for the refinery industry is to adapt the quality of its oil products to new national clean air standards adopted in 1993. These regulations are aimed at alleviating the pollution problems of Italy's major cities. The decree stipulates that the benzene content of petrol must be reduced by 2.5 per cent and that of aromatics by 33 per cent. The sulphur content of diesel must not exceed 0.2 per cent and that of heating fuel oil 0.3 per cent. Another impending challenge, particularly for the smaller refiners, will be to increase output of unleaded petrol, which until 1992 represented only 14 per cent of total petrol sales in Italy.

Central Europe

The energy deficit of this region is particularly severe in the oil sector, explaining the limited share of this fuel in the market-place and the heavy dependency on imports. Until 1990 under the Communist regime, supplies provided by the Soviet Union were sold to CMEA partners at preferential terms. For example, before 1989 East European countries paid less than $7.5/barrel for Soviet crude (Richter, 1992). End-user prices for refined products were also heavily subsidized, a policy that for many decades was believed to promote industrial development and social welfare. As is the case in Western Europe, oil was phased out of electricity generation and attention was concentrated on meeting the demand in transport.

Governments in Central Europe were relatively quick to reform the oil sector after the Communist regimes had collapsed. Oil enterprises were transformed into financially independent joint-stock companies. Some have been privatized, others remain state-owned, but they generally operate in a competitive market environment. Price regulations have for a large part been lifted

but there are still cases where governments are involved in pricing issues for
oil products, notably in Poland.

3.5. Current Issues

From Supply Security to Market Liberalization

One important preoccupation in European energy policy has been, and will
continue to be, Europe's vulnerability to supply shortages. At present international crude oil markets are sufficiently well supplied, while breakthrough
improvements in oil exploration and production technologies have removed
earlier apprehensions about the long-term scarcity of oil. Western Europe has
also strengthened its capacity to withstand supply contingencies through oil
emergency storages and international cooperation.

The remaining supply concern is that of ensuring that European oil production, primarily in the North Sea, be maintained at its present level for as long as
possible, taking into account the fact that the region is already a mature oil
province. This concern is balanced by the internationalization of European oil
companies and by the establishment of multinational treaties improving the
conditions for these companies to operate in other countries, for example those
which had signed the Energy Charter (see Chapter 8).

Market liberalization has also been a central element in oil policy, aimed at
securing the availability of crude and oil products at competitive prices. Most
trade barriers in oil-product markets have been removed across Europe. Exposure to international market forces has also had significant impact on the
European refinery industry by forcing the sector into a badly needed capacity
rationalization process that today is almost completed.

The Oil Market and the Environment

The environment is the new agenda for the West European oil industry. During
the last decade oil refineries were forced to adopt processes reducing the
sulphur content of oil products, a requirement that undermined the competitive
position of fuel oil in stationary applications, though it did oblige oil companies to strengthen the position of light and middle distillates in the transport
sector. In principle the European refinery sector would now be well equipped
to satisfy future regional market needs, were it not for new regulations further
enhancing environmental quality standards for oil products. At present the
most relevant areas in which environmental regulations affecting the oil

industry are being adopted, or are under debate, involve abatement measures for SO_2, NO_x, volatile organic compounds (VOCs), and CO_2. While the implementation of measures for reducing CO_2 emissions have not been conclusive, a subject that is discussed in Chapter 7, there is a mounting consensus to tighten the regulations to abate sulphur emissions. In June 1994 over thirty countries signed the second UN ECE sulphur protocol, which sets individual SO_2 emission targets for the years 2000, 2005, and 2010. All EU governments having signed the protocol, the European Commission now seeks to accelerate work on a 'liquid fuels directive' specifically designed to set sulphur limits for fuel oil, bunker oil, gasoil, jet fuel, and diesel. It is also working on another directive to reduce the sulphur emission of combustion plants under 50 MW capacity, which were not covered by the Large Combustion Plant Directive of 1988.

Estimates for the investments needed by the refinery industry to comply with new environmental regulations indicate that compliance with the directives under debate could force many European refineries into closing some of their plants. The European oil industry organization Europia argues that previous directives already cover the concerns of the Community in a satisfactory way, and that decisions should be left to member states about the specific measures they should adopt to abate sulphur emissions at the national level.

Structural Changes in the Oil Market

This sector has undergone a radical transformation since the 1970s. Then, the industrial structure was dominated by the vertically integrated multinational majors and the national oil companies often established as instruments of public policy. In the following decades both types of enterprises and the entire European oil market have changed dramatically: The majors have lost control over the long chain of activities from production to distribution. Governments have sold, or are in the process of selling, major assets in what used to be 'their' national oil companies. Barriers to trade in oil products and public price controls have largely disappeared, opening up to increasing international competition.

In short, the oil market in Europe has developed from an intensely politicized arena, closely watched by public and private actors, into a market-place that still attracts political attention, but is left primarily to the interplay of market forces that operate across national borders. In this process the role of governments has gone through a profound change. In the 1970s the prevailing social contract bound companies and national authorities together: the latter provided political support, in some countries overt trade protection, so that the former could guarantee secure supplies of oil. This was a tacit bargain in the relationship between the majors and their home governments and a public commitment in the countries with national oil companies.

In both cases this social contract has been replaced by the logic of the market-place. Governments across Europe now see their interests best served by removing barriers and favours previously given to national companies. More competition is assumed to provide more efficient production, while maintaining a reasonable level of security of supply. Even if most governments have adopted an arms-length attitude to the oil industry, they retain political objectives of major importance to the future of the companies operating within this part of the energy sector.

The Tax Base

Across Europe, public authorities have without exception taken advantage of the special features of the oil industry. It manufactures products which most users, especially in the transport sector, can hardly do without and are therefore willing to pay a high, and growing, price for. It is a highly competitive industry with strong pressures on commercial margins in every part of the business. It is based, with few exceptions, on imported raw materials. In consequence, it provides ample opportunities for increasing public taxation. As a consequence European governments have managed to transfer a significant share of the oil rent from the industry to the treasuries, while the producer countries have not increased their share of the oil rent. This has made the oil market a major source of public revenues, which is by now the most important part of the social contract as concerns this subsector of European energy.

Environmental Concerns

As we have described, the oil industry is facing escalating environmental regulations. Refineries that cannot cope with higher environmental standards, will be forced to close, following the trend that began in the 1980s. The major companies seem to want more freedom of manœuvre and less detailed public intervention in the field of the environment, but politicians and voters are uncertain of how to deal with the serious ecological problems emanating from the use of oil products, in particular from transport in urban areas. By the middle of the 1990s there is no stable social contract between the industry and the public on these issues, which leaves the future open to diverging trends, to which we shall return in Part II.

Security of Supply

Given the importance of oil for modern transport systems, governments will always take an interest in these matters. As of now, this is in practice confined to stockpiling measures, normally in close cooperation with industry, and international collaboration, primarily within the IEA. Should a new international supply crisis occur, however, direct government intervention cannot be ruled out. In the longer term this could also affect the structure of the industry, perhaps even re-establishing a social contract akin to the 1970s—major companies acquiring protection from competition in return for assurances of secure supplies.

4
Energy-Sector Developments: Coal

Throughout most of the period since 1970 European hard-coal production has suffered major economic losses. Virtually all German and French production has been, and is, uneconomic, and a major part of the production in UK and Spain would not have survived without protection. The economic viability of the Central European coal industry is difficult to assess, but a major part of the production would probably not be sustainable in a market economy and, as has been shown over the past few years, significant volumes of production are not even marketable despite considerable financial support.

The future of the European coal industry is thus highly dependent on governmental policies. Various factors are important in this respect, some of which lie outside the domain of energy policy. Employment and social considerations, together with concerns for the balance of payments and import dependence or energy security, are arguments favouring continued aid to the coal industry. Concerns about public finances, the environment, and energy-sector reforms represent pressures on current policies. In order to assess the importance of these factors for future coal production, it is necessary to look at the past policies and structure of the coal sector in the major coal-producing countries. Future developments will be determined largely by historical policies towards the coal industry, including the national characteristics and strength of vested interests.

A policy supporting national coal production is not necessarily to encourage the use of coal. Protection of indigenous coal also keeps less expensive international coal out of the market. It is therefore important to distinguish between policies directed towards coal production and policies that have an effect on coal consumption. This chapter deals primarily with policies in support of coal production. Demand-side measures are also to some extent discussed in this chapter and subsequently in Chapter 6, on electricity.

4.1. Historical Background

Western Europe

For two-thirds of the twentieth century coal was the dominant energy source in Western Europe and the coal sector was the principal action ground for energy policy. Not only did coal dominate in electricity generation and industrial processes, but it was also the principal energy source for space heating and cooking in the household sector, often through the manufacture of town-gas. It was not until 1966 that oil replaced coal as the most important energy source in Western Europe. The share of coal in total primary energy supply underwent a radical decline, from 85 per cent in 1950 to 34 per cent by 1970. West European coal consumption peaked in 1956, at about 575 million tonnes of coal equivalent (mtce), falling to 400 mtce in the early 1970s. Coal lost market shares to cheaper and more convenient oil and natural gas—in industry and electricity generation as well as the residential and commercial sectors. After 1974 coal consumption stabilized as the increase in demand for coal for electricity generation compensated for the continued decline in the industrial market. However, the coal share in total energy supplies continued to fall, down to 20 per cent by 1994.

The reduction in coal production was steeper than for consumption, as more demand was gradually directed towards imports. Prior to 1960, coal trade had mostly been intra-regional and land-based. In the 1960s and 1970s an international market in sea-borne coal trade emerged, leading to an increase in imports of coal from non-European sources. From 1960 to 1970 net imports grew from 35 mtce to 54 mtce, with a further increase to 150 mtce in 1994. This development has made Western Europe one of the major markets for international coal trade.

All West European coal-producing countries, with the exception of Spain, saw major reductions in production and demand prior to 1970 (see Fig. 4.1). In the two leading coal-producing countries, Germany and the United Kingdom, production fell by 60 and 75 per cent from 1960 to 1992. Their share of total coal production, however, has increased as other countries experienced an even steeper decline in production. France had its production reduced by 85 per cent between 1960 and 1994. In the Netherlands production was discontinued in 1975 and Belgium closed its last pit in 1992.

Central Europe

In Central Europe coal remained the dominant fuel throughout the Communist period. The structure of the electricity sector, major parts of industry, and even

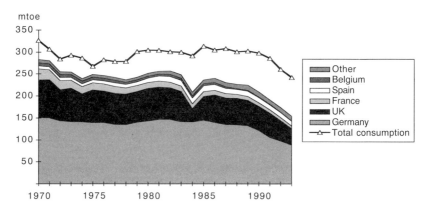

Fɪɢ. 4.1. Coal production and consumption in Western Europe
Source: IEA (1994*a*).
Note: 1 mtce = 0.7 mtoe (million tons of oil equivalents).

the residential sector heating system were adapted to a high share of coal. In electricity generation, 60–80 per cent of the energy used was generated by hard coal or lignite. In industry, the emphasis was on developing energy-intensive sectors with a high share of coal. In the residential sector, most of space-heating needs were covered by district heating, supplied primarily from coal.

Although the share of coal in Central Europe's total energy consumption has declined somewhat as a result of the recent economic crisis and restructuring, coal remains the dominant fuel in the energy balance, as it was in Western Europe in the 1950s and part of the 1960s. In Poland, coal accounted for 75–80 per cent of primary energy consumption throughout the 1970s and 1980s. In Czechoslovakia, the share was 75 per cent in the 1970 but shrank gradually to 55 per cent by the early 1990s as nuclear power and new supplies of natural gas became available. In Hungary too, coal was replaced by natural gas and nuclear power, partly because geological conditions did not permit growth in coal production in line with expanding energy requirements.

Czechoslovakia and Hungary are land-locked countries with only modest reserves of hydrocarbons. Their principal source of external energy supplies is Russia. This explains the political emphasis given to the development of indigenous coal, despite sometimes difficult geological conditions and poor quality (high sulphur and ash content). Poland, on the other hand, has a better reserve basis, but its coal sector has been developed far beyond the economic optimum. Therefore the problems of retrenchment and restructuring of the coal sector are no less challenging in Poland than elsewhere in the region.

4.2. Coal Policies in Selected European Countries

Germany

Germany is a major producer of both hard coal and lignite. Most of the lignite is produced in the new Länder but there is also notable production in the old Länder (West Germany). Here the lignite production is for the most part viable without state subsidies, while hard-coal production is by and large unprofitable. Coal policies in Germany have primarily been directed towards support of hard-coal production.

Hard Coal

German hard-coal production declined steadily from 150 mtce in the mid-1950s to 55 mtce in 1994. The reduction in the number of workers over the same period was even more dramatic: 600,000 were employed in 1955, as against 102,000 in 1994. From the early 1960s to the mid-1970s there was a sharp decline in German coal use outside electricity generation (see Fig. 4.2). Coal had been widely used for town-gas production, which was replaced by natural gas. Also, the demand for coal in industry, in particular the iron and steel industry, has declined. Higher end-use efficiency and a decline in the

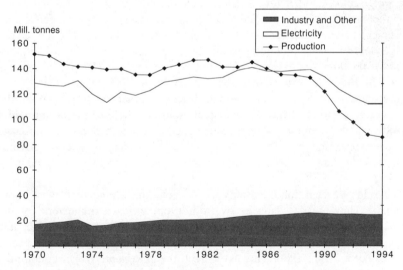

Fɪɢ. 4.2. Coal production and use in Germany
Source: IEA (1994*a*).

production of raw steel are the main factors explaining this. Coal use in electricity generation, on the other hand, increased slightly from the late 1960s until the early 1980s, and has since remained almost constant at about 40 mtce per year. German coal production has declined faster than domestic demand and net imports have increased since the early 1980s. Imports are, however, still relatively small in terms of the total German market.

German coal mines are privately owned. There are two main coal producers, Ruhrkohle and Saarbergwerke, located in the Federal Länder of Rheinland-Westphalia and Saar respectively. These two Länder produce almost all German hard coal and have had to face the social and political difficulties related to reduced employment in the coal industry. Policies have been directed at limiting the inevitable decline in employment by restricting the access to imported coal and by providing direct and indirect subsidies to the coal mines. Political pressure for measures to even out the decline in production and employment have been supported by the concentration of coal production in the Ruhr region.

Both coking coal for steel production and steam coal for electricity generation are protected from competition through a network of legislation and private agreements. In addition, there is direct and indirect price subsidization, including subsidies for investments and other production-related expenses. Two systems of production support have been put in place: one for the iron and steel sector, another for electricity generation.

For coal use in the iron and steel industry, sales are guaranteed by an agreement between the coal producers and the steel industry. The agreement obliges German steel producers to purchase practically all their coal requirements from domestic sources. The first agreement, the *Hüttenvertrag*, was signed in 1969 and extended to the year 2000 in 1985. Coal prices in this contract are kept close to international prices in order not to jeopardize the competitive position of the German steel industry. The coal mines are paid a subsidy which is equivalent to the difference between their production costs and the sales price. Subsidy funding is shared between the Federal government and the Länder governments, mainly those of the two coal-producing Länder.

For sales of coal for electricity generation, the arrangements are more complex. They are based on a law from 1965 setting a framework for the use of German-produced coal in electricity generation. The specifics of the law are embodied in an agreement between the coal mines and the electricity utilities dating from 1980. This agreement, the *Jahrhundertvertrag*, commits the utilities to purchase increasing amounts of German coal, while imports of coal from non-EC countries are restricted under a quota system. Basically, imports are allowed only for coal requirements in excess of the purchase obligations for German coal. Purchase obligations here are more strict than for the iron and steel industry. The latter is only obliged to purchase its actual needs from German mines, whereas electricity generators must purchase the volumes stipulated in the agreement, even if these may be in excess of their needs, until volume commitments are renegotiated. Purchase obligations of the electricity utilities have been reduced in several rounds. Less growth in electricity demand than originally expected and additions of new nuclear-power generation capacity have limited the utilities' capacity for absorbing German-produced coal. The result was stockpiling of coal in the second half of the 1980s and the early 1990s.

Germany's electricity utilities have paid a coal price which covers the costs of production, net of subsidies. Production costs are estimated to be about three times the average delivered price on world markets (see Table 4.1 and Box 4.1). To compensate utilities for the high costs of steam coal and to share the effects on electricity prices between utilities, a coal levy was charged on all electricity sales ('the Kohlepfennig')—8.5 per cent in 1994. The revenues from the levy were paid to utilities using German-produced coal. Refunds were granted to bring the price of coal to the level of heavy fuel oil for 22 million tonnes and to the price of imported coal for 11 million tonnes (although with a ceiling for the subsidy). The remainder of the utilities' coal purchases was not subsidized, which means that utilities pay the actual production costs (Heilemann and Hillebrand, 1992).

Largely due to this coal policy, electricity prices in Germany have been relatively high, with considerable disparities within the country. This was workable only because of the monopoly and area protection enjoyed by the German electricity utilities. German utilities with relatively low fuel costs were excluded from competing with other utilities that have relatively high costs. The market is also closed for competition from electricity producers outside Germany.

In December 1994 the Federal Constitutional Court ruled that 'the Kohlepfennig is inconsistent with Germany's financial constitution since its objective—energy security—is of general interest and therefore should not be placed as a special burden on electricity consumers'. The 'Kohlepfennig' will be replaced by a scheme of direct support for the coal industry through Federal Länder budgets. As a result electricity prices will fall.

Total price and production subsidies (PSE) amount to some 8–9 bn US$ annually, corresponding to 170–90 US$ per tce of coal produced (see Table 4.1), which will be funded from the public budgets. Production costs are almost 165 US$ per tce. For comparison, the price of imported steam coal

Table 4.1. *Financial aid to coal production in Germany, 1987–1994 (US $m)*

	1987	1989	1991	1994
Total producer subsidies (PSE)	5,919	8,500	9,432	8,025
Aid not benefiting production	5,214	5,636	6,982	n.a.
Total aid	11,133	14,136	16,414	n.a.
PSE/production (US$/tce)	72	187	189	172

Source: IEA (1993), and IEA (1995*c*).

was about 40 US$ in Rotterdam in 1994. Including transportation and other costs, imported coal would cost electricity utilities in the Ruhr area about 55 US$ per tce. If other subsidies are added, total subsidies come to about 12–14 bn US$ or some 225–50 US$ per tce of produced coal. Other subsidies are mainly public payments to cover the deficit of the miners' pension funds, subsidies which are independent of current production levels (see Box 4.1).

The inability of German coal to compete with internationally traded coal is clearly set out in the 1990 report of the German Coal Commission. According to the report only three pits, representing 15 per cent of indigenous production, had average production costs below DM 225 (US$ 135) tce. Moreover there is scant potential for cost reductions, given the rather poor geological conditions for coal extraction in Germany. It therefore appears likely that production costs for even the best of the German mines will remain two to three times above the costs of purchasing coal on the international market.

The productivity of 710 kg per man-hour in German coal mines results in a price and production subsidy of almost 80 US$ per man-hour worked in the mines. Production-dependent subsidies are thus significantly higher than miners' wages. It would cost less to terminate all production and pay the miners' full salaries than to continue production.

Despite its high costs, Germany's coal support system is partly kept in place by public ownership of some key coal-consuming electricity utilities. Municipalities and Länder authorities of the main coal-producing region, Rhineland-Westphalia, have considerable influence on the overall strategies of the local electricity utility (RWE). Although the utilities publicly argue for a reduction of coal prices to world-market levels, securing employment in coal mining still has priority in Länder politics. Rhineland-Westphalia is also the most populous of the Bundesländer and has a central role in Federal politics. All these concerns act together to preserve Germany's coal policies.

The policies have, however, been under political pressure, both from the EU Commission and internally in Germany, particularly from the Liberal

Box 4.1 *Public support for the coal industry*

Since 1988 the International Energy Agency (IEA) has published data on public support to the coal industry in its member countries. The data are compiled and presented using a methodology outlined in IEA (1988). Financial support to the sector is divided into two categories:

> *Assistance to current production*, using the Producer · Subsidy Equivalent calculation (PSE)
> Assistance not benefiting current production

PSE is meant to reflect the support provided to current domestic production of coal that the industry would be expected to cover in a competitive situation. PSE normally consists of two main elements: (i) direct and indirect financial aid to current production, including investment grants, deficit grants, and support to miners' pension funds; (ii) price support. The first element generally takes the form of direct state payments, whereas price support can include both direct budgetary support and excess payments by coal consumers. Excess payment is calculated as the actual price paid by domestic consumers and a reference price reflecting the costs of purchasing comparable coal qualities on the international coal market and under similar contractual conditions. Price support is for most countries the largest element of PSE. One important part of this support has been special sales arrangements between coal producers and electricity generators, particularly in the case of Germany and United Kingdom. The two countries have argued that this should not be counted as a coal subsidy, since long-term arrangements for local coal deliveries are needed to fulfil the utilities' own supply obligations.

Assistance not benefiting current production, which for most countries is as significant as PSE, normally takes the form of direct public transfer. The largest items within this category are liabilities related to pension funds to former employees, support to environmental clean-up from past mining activities, and support designed to speed contraction of the industry.

The distinction between PSE and assistance not benefiting current production may appear clear, but it could be argued that PSE understates the actual subsidization of current production. For example, current costs and support requirement can be low because of previous debt write-off, counted as assistance not benefiting current production. Likewise current costs can be kept low if there is an understanding that public support will be granted in some future year to obligations like pensions or environmental clean-up. (See Radetzki, 1994b.)

Source: IEA (1988)

Party. The EU Commission argues that the subsidization schemes should be directed at a restructuring and rationalization of the coal-mining industry. The Commission approved in 1992 the *Jahrhundertvertrag* for the period up to 1995. Subsidies to domestic energy resources may be accepted for a maximum of 20 per cent of total primary energy demand by 1995 and 15 per cent by 2000. By using this formula, the Commission implicitly accepted that security-of-supply concerns justify maintaining German coal production.

Lignite

Lignite production in Eastern Germany (the new Länder) was 250 million tonnes in 1990, i.e. more than half of all European lignite production. After unification, production has been reduced sharply. The government expects lignite production in the new Länder to stabilize at less than 100 million tonnes (IEA, 1993). The rationalization of lignite mining is intended to make production fully competitive, with no need for public support, though some of the new Länder have decided to give transitional subsidies to power plants based on local lignite (IEA, 1992a). In order to guarantee the market for electricity generated from lignite, an agreement between the former East German and West German governments concluded in 1990 (*Stromvertrag*) obliges distribution companies in the new Länder to purchase a minimum of 70 per cent of their electricity needs from a specific transmission company. This company, VEAG, owned by the West German utilities, will in turn buy lignite from a consortium, also owned by some of the dominant West German utilities. The *Stromvertrag* was contested by the distribution companies and the major towns, who wanted more freedom to generate power locally and feared that the electricity sector in the new Länder would become as mono-polistic as in West Germany. A compromise was reached which retains the obligation to purchase 70 per cent of power needs from VEAG.

United Kingdom

The British coal industry was nationalized in the late 1940s. During the war, coal production had been strictly government-controlled as a strategic sector for supporting electricity generation and industry, in particular the iron and steel industry. The Labour Government of the first post-war years aimed to keep control of the key sectors of the economy, including energy. This partly reflected a long-term strategy of establishing a mixture of state and private ownership of industry.

Since the early 1960s, there has been a fundamental change in the structure and level of coal demand in the UK (see Fig. 4.3). Total coal demand has been

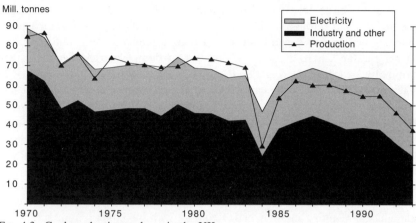

FIG. 4.3. Coal production and use in the UK
Source: IEA (1994*a*)

halved, with the most rapid decline taking place during the 1960s and into the early 1970s. In the 1950s and 1960s coal was still an important fuel for household space heating and for production of town-gas—in both of which it was later replaced by other fuels, in particular by natural gas. State-owned British Coal lost in this process about one-third of its market. Later, industrial coal demand declined in general industry, where gas and oil gained market shares, and in the steel industry. Coal use in electricity generation, on the other hand, increased until its peak in the early 1980s. UK coal production is primarily geared to satisfy the domestic market. There was a balance between production and domestic demand until the miners' strike in the mid-1980s; since then, net imports have gradually increased, representing one-fifth of coal consumption in the early 1990s.

From the late 1970s and early 1980s, power generation became the dominant user of UK coal, accounting for about 80 per cent of total consumption. In order to secure the market in the face of increasing competition from cheaper internationally traded coal, the previous informal links between the coal and electricity sectors were made more explicit. British Coal and the Central Electricity Generating Board (CEGB), both state-owned enterprises, entered into framework agreements guaranteeing coal sales for electricity generation. These agreements, known as 'Joint Understandings', were made under strong pressure from the government. In the first Joint Understanding from 1979, the coal price was set in line with British Coal's average operating costs, and CEGB—on a best-endeavour basis—committed itself to purchase coal volumes somewhat higher than current use at the time.

All of the costs not covered by revenues from coal sales were financed by the government through a direct subsidy, the deficit grant. The Joint Understanding was revised several times, in particular reducing the volume commitments for the CEGB because of lower growth in electricity demand than originally expected. During the 1980s, CEGB paid more for UK coal than international prices. In the early 1980s this surcharge was about £3 per tce, which increased to more than £10 per tce (IEA, 1988) as the price for delivered UK coal to power generation remained almost constant while the price of imported coal declined.

UK coal policies were drastically reversed after the Conservative Thatcher Government came to power in 1979. Subsidization of the coal sector was to cease, in line with the government's general programme of reducing public expenditure and state involvement in business of any sort. Mine closures were announced, resulting in an extended conflict with the miners' trade union and also with the TUC, the umbrella organization for the trade-union movement. The subsequent miners' strike in 1983/4, which lasted for almost a year, developed into a confrontation with the government not only about the future of British coal mining, but also about the role of trade unions in state-owned enterprises and in the economy in general. Formerly, no major decisions affecting employment in coal mining had been taken without obtaining consensus on broad policy lines with the trade unions. For the Thatcher Government, the miners' strike became a test of the government's power to destroy the hegemony of the trade unions and to carry through its programme of reforming the economy.

The CEGB managed to cope with the shortfall in supplies of UK coal during the strike by drawing on its own coal stocks, by stepping up imports of coal, and by a significant increase in oil use in power generation. Oil-fired reserve capacity was brought back into operation. The strike ended in defeat for the miners' union, and this in turn paved the way for a programme of mine closures and reductions in state subsidies.

Table 4.2. *Financial aid to coal production in the UK, 1987–1992 (US$m)*

	1986/87	1990/91	1993/94
Total producer subsidies (PSE)	2,053	2,087	336
Aid not benefiting production	1,806	769	891
Total aid	3,859	2,856	1,227
PSE/production (US$/tce)	19	22	5

Source: IEA (1995c).

Out of 170 pits, 120 were closed between 1975 and 1992; employment fell from 220,000 to 54,000. In the same period productivity (tonnes per employee) more than doubled. Despite these improvements, the contraction of employment and coal production accelerated after 1992. By the end of 1994 employment was down to 11,000 workers, with indigenous production for 1994 estimated at 42 million tonnes. The immediate cause of this reduction was the privatization of the electricity sector and the 'dash for gas', to cite Newbery (1994). The competitive advantage of new and efficient combined-cycle gas turbines (CCGT), together with the structure chosen for the privatized electricity sector, has led to major investments in gas-fired capacity which eventually will displace some 30 million tonnes of coal (see Chapter 6). Consequently, in negotiating purchases from British Coal beyond 1993 the privatized electricity enterprises, National Power and PowerGen, were only willing to contract for 30 million tonnes of coal from the mid-1990s. This is less than half the quantity sold to the power sector under the old contract.

Following British Coal's announcement in late 1992 of a programme of rapid and extensive mines closures, the UK Government undertook a review of the industry. This review—like that undertaken by the House of Commons—was aimed at defusing considerable public criticism of the mines closure programme. Following the review, the government recommended that twelve of the thirty-one mines originally scheduled for closure should be 'reprieved'; closures should be temporarily suspended, and these mines should be allowed a trial period during which their ability to find a market outlet for their production should be assessed. In support of this, the government promised to reintroduce state production subsidies, but only for these mines and only until they could be privatized. This programme had no apparent success. Most of the 'reprieved' mines ceased production during 1993 and 1994. The most important problem was not resolved, the absence of new demand for coal for power generation:

Without a parallel creation of breathing space in the market to allow for a growth in the share of subsidised coal, the effect of any subsidy was likely to remain minimal, particularly when the extent of subsidy remained unspecified and short term. Under such circumstances, it seems clear that the interests of the generators were best served by continued diversification away from coal (Bruce, 1994).

British Coal's contract with the electricity companies runs to 1998. The future of coal production beyond this will depend on the ability of the coal industry, which was privatized in 1994, to compete with gas and imported coal for power generation. According to British Coal, production costs could be reduced to £26–30 per tonne by 1998, compared to £40 per tonne recorded in 1992. This would make 45 million tonnes competitive at inland stations against imported coal. But, as conceded by British Gas, 'the industry would need some degree of price protection for the next 3–4 years' (Baker and Rendall, 1993) in order to maintain production capacity and restructure the industry.

The privatization of the electricity industry and the deregulation of the gas industry have accelerated the decline in coal use in power generation in the UK. Private investor interests in the electricity sector favour gas-fired plants for new generation capacity because of relatively low gas prices, in combination with the lower capital requirements compared to coal plants. In addition, the privatized regional electricity companies have obtained the right to construct their own power plants within certain limits, for which the preferred fuel is gas. This 'dash for gas' in electricity generation risks reducing coal production in the long term at lower levels than is economically justified. Reopening a closed coal mine often entails prohibitive costs.

France

Compared to other European countries France has since the 1950s done little to support indigenous coal production. The 1960s saw a major shift in energy consumption towards cheap and convenient oil. Oil was readily available from the Middle East and from Algeria, which initially was considered a domestic resource. The subsequent nuclear programme and contraction of coal demand in industry and the residential sector led to a 70 per cent reduction in coal production from 1970 to 1992. During the 1970s the government allowed coal imports to increase. Coal ports and other infrastructure were built to support the increase in demand for foreign coal. Net imports peaked in 1980 at 39 mtce, far above any other European country. After 1980 coal imports declined as more nuclear capacity came into operation, but France remains the largest coal importer in Europe.

All coal production in France is in the hands of the state-owned Charbonnages de France (CdF). The main customer is Electricité de France (EdF), also

Table 4.3. *Financial aid to coal production in France, 1987–1992*

	1987	1989	1991	1992
Total producer subsidies (PSE in US$m)	373	152	143	165
PSE/production (US$/tce)	29	14	16	21

Source: Piper (1994).

state-owned, both for direct coal sales and for sales of electricity produced by CdF in its own power plants fired with coal directly from the pit-head. An agreement from 1987 between EdF and CdF stipulates the annual coal volumes and CdF-produced electricity which EdF is committed to purchase. Coal is sold at prices close to world-market prices and electricity at the same price as other auto producers obtain from EdF in France. CdF operates with losses covered by government subsidies.

The policy agreed between CdF and the French Government is to phase out coal production by the year 2005 at the latest, but in step with voluntary retirements from the workforce. Actual decisions on mine closures are therefore taken as a function of developments in the workforce. This gradual approach and the emphasis on voluntary retirement has succeeded in avoiding any major social unrest in the coal-producing regions.

Spain

Unlike other coal-producing countries in Western Europe, Spain increased its coal production between the 1970s (10 mtce) and 1985 (19 mtce) and has only contracted moderately thereafter. Growth in demand for coal, particularly in the electricity sector which makes up 80 per cent of coal use, required an increase in coal imports as well. From 1970 to 1994 net imports of coal rose from 2.3 mtce to 11 mtce. The growth was particularly strong in the late 1980s; Spain has in recent years invested heavily in infrastructure for coal imports (IEA, 1994*a*).

The system of protection and subsidization for Spanish coal production is similar to that in force in Germany until 1995. Import restrictions protect against internationally traded lower cost coal, long-term agreements guarantee sales of coal to the electricity sector, and a levy on electricity sales, including electricity generated by other fuels, is used to compensate the electricity utilities for a part of the additional costs. Major parts of Spain's coal production

Table 4.4. *Financial aid to coal production in Spain, 1987–1992 (US$m)*

	1987	1989	1991	1994
Total producer subsidies (PSE)	533	596	708	962
Aid not benefiting production	622	n.a.	n.a.	n.a.
Total aid	1,058	n.a.	n.a.	n.a.
PSE/production (US$/tce)	53	34	39	57

Source: IEA (1993) and (1995*c*).

are uneconomic, but are maintained because of concerns for local employment. A restructuring programme was imposed through a ministerial decree in 1990. Aid was provided to close down several non-viable mines and to rationalize others. Of the seventy-one privately owned mines, which make up some 85–90 per cent of total coal produced, fifty-three were affected by the programme. At the end of 1993 some 3 million tonnes of production capacity had been closed and the workforce was reduced by some 6,000 workers.

Belgium

Coal is the only indigenous energy resource in Belgium. Coal production was therefore maintained at a level close to 15 per cent of total energy supplies during the 1970s, despite major losses in the mining industry. The geological conditions for coal production in Belgium are poor in comparison with other European countries. In the 1980s productivity in coal mining was about two-thirds of the Western European average (IEA, 1988). Financial aid during the 1980s escalated both because productivity gains were not made in domestic production and due to lower coal prices on the international market. By 1989, Producer Subsidy Equivalents stood at US$ 118 per tce, approximately 20 per cent above the level in Germany. A decision gradually to shut down indigenous production was taken at the end of 1986. The last mine, Zolder, ceased production in September 1992 with no major social problems (IEA, 1993).

Poland

Historically coal has held an important position in the Polish economy. In 1990 the sector accounted for 10 per cent of gross domestic product and a similar share of the country's export earnings (IEA, 1991*c*). Coal exports have declined since 1990, but are still important for the economy. In 1992 they totalled $900m, or 7 per cent of total Polish exports of goods. Employment in

Table 4.5. *Financial aid to coal production in Belgium, 1987–1992 (US$m)*

	1987	1989	1991	1992
Total producer subsidies (PSE)	324	184	77	50
Aid not benefiting production	920	141	147	137
Total aid	1,245	325	224	187
PSE/production (US$/tce)	81	97	76	76

Source: IEA (1993).

the coal industry was 305,000 persons (Radetzki, 1994*a*), i.e. 2 per cent of the total labour force. Employment has declined, from 430,000 in 1987.

Under the Communist system, coal prices were set at artificially low levels. Capital costs were kept low, while labour costs were high, including a wide range of social services for workers and their families—housing, schools, hospitals, education. In the process of reform, coal prices on the Polish market for power generation and industry have been aligned to international prices. In the current situation of a surplus supply of coal, the yardstick for the domestic price is the export price netted back to the mines by deducting rail transport costs of about $10/ton to a coal export harbour. In 1993–4, Polish coal prices, from the mine, were about $25/ton compared to an international price of $45–50/ton CIF Northern Europe. This is a relatively low domestic price compared to the international level. Comparisons are, however, difficult to make. Coal sold for power generation has an average ash content of more than 20 per cent, a quality not traded on the international markets, where steam coal has a typical ash content of less than 10 per cent.

For the existing mines, most production is currently uneconomic. Of the total number of seventy hard-coal mines, only four produced a profit in 1992 and the total operating deficit was in the order of US$ 800–900 m. The market price for coal is lower than the total costs of production. As a result, the financial situation of state-owned mines is deteriorating, and they generate insufficient revenues to cover depreciation. A decision has been made to close eighteen pits, but also to open seven new ones. In addition to the economic difficulties of maintaining current production, a reduction of the environmental impact of coal mining—in particular from salination of rivers and improvements in the quality of marketed coal—will require investments that will further burden the mines' financial situation (IEA, 1995*a*).

Closing the least efficient mines would raise productivity, improve the market balance, and raise coal prices. Consequently, the financial situation of the remaining coal mines would improve. Labour productivity in Polish coal mines is some 35 per cent lower than in Germany and 75 per cent lower than in the UK, which has the highest productivity level in Europe. Closures would, however, accelerate the reduction in employment in the mines.

Coal will retain an important share in Poland's energy supplies, but there are no prospects for any important increase in domestic coal demand which could halt mine closures. The use of coal in energy-intensive industries, in particular in iron and steel, is declining because of restructuring and closures. Higher efficiency in the district heating systems and in residential space heating should over time reduce coal demand. In addition, the government plans some diversification of Polish energy supplies in order to reduce the dependence on coal and thereby weaken the political power of the miners' unions. The potential for exports is limited by transport costs. Polish coal cannot compete with alternative supplies in locations served by large transoceanic ships. It will, however, remain competitive in Baltic and North Sea harbours, which can only receive small ships (Radetzki, 1994*a*).

Closures of the least economic parts of Polish coal production could probably create an economically viable coal-mining industry, but the lack of alternative employment for the miners risks freezing the situation politically. Protection of employment in the mines could be given higher priority than rationalization of coal production. Such a development would mean a repetition of the experience of the Western European coal-producing countries, which, faced with declining coal demand and increased competition from imported coal and other fuels, protected and subsidized their domestic coal industries.

The Czech and Slovak Republics

As in Poland, coal dominates energy use in the former Czechoslovakia. In the 1970s, coal accounted for 75 per cent of total energy use in the country (the two republics as a whole), but the share has in recent years declined to 55 per cent. There has been a decline in the use of coal since the mid-1980s due to reduced industrial coal consumption. The Republics are self-sufficient in coal, and production has decreased in line with falling demand.

About 80 per cent of coal production is low-quality lignite. Ash content averages 25 per cent and sulphur content 1.7 per cent, which causes severe environmental impacts. There are also environmental problems linked to the many open-cast mines, which use extensive tracts of land. The costs of this land use can be considerable, and are not fully reflected in mining cost figures.

Production costs are generally high and productivity is low, both in underground hard-coal mines and in open-cast pits. In both productivity is about half the Western European average (IEA, 1992*b*). Average production costs in the period from 1981 to 1985 varied by a factor of 5. The 45 per cent most efficient had a cost of about 13 $/tce and the least efficient 10 per cent had costs higher than 45 $/tce. A large part, possibly all, of hard-coal production cannot compete with imported coal in the future. There are, however, uncertainties about cost reductions that can be obtained through rationalization. Open-cast lignite mines with associated power stations have better prospects, as these plants often have operating costs lower than the full costs of any alternative.

In the Czech Republic, by far the more important coal producer of the two countries, there has since 1989 been a restructuring of the coal sector. Six joint-stock mining companies have been created—three for hard coal and three for lignite mining—which will be majority-owned by private investors. It is government policy to grant subsidies to mining only as a transitional measure. Employment was reduced from 195,000 to 105,000 between 1989 and 1993, and the thirteen least profitable mines have been closed. The cut-backs have concerned both lignite and hard-coal production. In the process, labour productivity has been stepped up significantly, with a 36 per cent improvement in the hard-coal mines and 43 per cent in lignite (IEA, 1994*b*).

The restructuring of the sector is far from over. Only rough estimates are available of the level of production and employment which is economically viable in the long term. It is estimated (IEA, 1994*b*) that economically sustainable hard-coal production will probably not exceed 5 million tonnes a year (compared to 18 million tonnes in 1993) and lignite production could be profitable with 30 million tonnes a year (67 million tonnes in 1993). Such levels of production and further productivity increases would suggest that total employment in the sector should not exceed about 15,000 after 2000. The reduction in the workforce has so far taken place without major social friction, but the relatively easy ways of managing personnel reductions, for example by early retirement, are now largely exhausted. Further and sharp reductions in the workforce could therefore trigger social tensions and conflicts. Although considerable progress has been made, particularly in restructuring the coal industry, some of the most difficult political decisions remain to be taken if the Republics are to avoid permanent protection and public subsidies in order to preserve production and employment.

Hungary

Coal has been much less important for the Hungarian economy than in Poland and Czechoslovakia. Currently coal covers about 20 per cent of total primary energy use, compared to 24 per cent in 1989, and production has been declining steadily in recent years. A major part of the coal goes to power generation. In the 1970s, about half of total electricity generation in Hungary was based on coal. In the 1980s, new nuclear power capacity covered all of the growth in electricity demand and also made it possible to reduce the use of oil in power generation. Coal use for power generation has declined slightly.

Only about 7 per cent of the total coal production in Hungary is of hard-coal quality. The remainder is brown coal and lignite. As in the Czech and Slovak Republics, the underground coal mines are relatively old and both costly and dangerous to exploit. Production from these mines was maintained for local employment reasons. The restructuring of the coal-mining sector has resulted in a sharp decline in employment. Half of the workforce left the industry between 1990 and autumn 1994, leaving only 18,000 workers in the coal mines integrated with the electricity plants and 5,000 outside. Although the total number of job-losses is relatively small in a nation-wide context, the social consequences are very serious. Social problems have been exacerbated by redundancies in the metallurgical industries and by the planned closure of two uranium mines, which will affect employment in the same geographical areas (IEA, 1995*b*).

The financial situation of the coal companies, chronically poor even before the collapse of the Communist regime, has deteriorated sharply, mainly because of the conversion of state capital subsidies into debt, bad investments, overstaffing, and controlled coal prices. Following liquidation of previous state enterprises, the sector has undergone a radical restructuring. Some mines have been integrated with power plant companies, others have been closed, and a few continue to operate temporarily under special conditions. Some of the non-integrated mines are operated as private companies formed by redundant miners, who lease the equipment from the state's coal restructuring company at prices significantly lower than the economic value. The mines integrated with the power plant companies receive no direct state subsidies, but they do possess a protected market by having a monopoly of supplying the power plant to which they are linked financially. They may, therefore, receive subsidies through the electricity sector. It appears that production from some of the mines is maintained at a cost which exceeds the price of alternative fuels for electricity generation. This form of subsidization of coal production through the electricity sector is reminiscent of the system in place earlier in the UK and Germany. At the end of the day, it is the electricity customer who pays the subsidies. The extent of subsidization is, however, not known.

About half of current coal production is from open-cast mines, mainly lignite mines. This part of Hungary's coal production represented 7 million tonnes of production in 1993. In the long term only production from open-cast mines is economically viable, and then only for a part of current level of production.

4.3. Pressure for Change

For Western European coal production, major policy issues relate to the future of the protection and subsidization of domestic coal mining, and in particular German coal mining. Seen in isolation, maintaining the protection of domestic coal industries has no direct consequences on the other Western European economies. None of them have any coal export interests of importance. This is probably also one of the reasons why the EU bodies for decades have accepted a continuation of the subsidization and protection of national coal production in, for example, Germany.

In Central Europe, coal is the dominant energy source, especially in Poland, where the employment issue is also the most important. There are in Poland strong trade union pressures for preserving mining jobs and the trade unions have important political leverage. Practically all of the electricity generation is from solid fuels, so a prolonged miners' strike would paralyse the entire economy. Coal policy issues in Poland are therefore not only linked to energy policies, but have more far-reaching repercussions. The other Eastern European countries have a more diversified energy mix, and coal-mine employment is less important.

There are also more wide-ranging policy issues related to the protection of European coal production than those related to consequences for the coal markets. We shall in the following focus on the major issues that will shape national coal policies.

The Employment Argument and the Power of Coal Trade Unions

If the European coal industry was to be fully exposed to competition from international coal, a large part of the 450,000 workforce in the industry would be made redundant. The social consequences of closing down European mines are considered to be significant, and such a policy is vehemently opposed by coal-industry interests. Historically, the power of the coal industry, in particular the coal trade unions, was based on the strategic importance of coal for industrial production and electricity supplies. The unions have been well organized and have had considerable political influence. Indigenous coal was the principal fuel for electricity generation and a prolonged coal strike could consequently cause major economic damage. Today, however, this power basis has almost vanished in Western Europe. Indigenous coal now accounts for less than 20 per cent of fuel supplies, against 80 per cent in 1960. What remains is the social impact of pit closures, which is higher than for other sectors because coal mines often are located in areas with no other alternatives for employment.

The remaining employment problem, however, is small compared to the major retrenchment that has taken place since the mid-1950s. In Western Europe as a whole the workforce in the coal sector has fallen from more than one million in 1955 to less than 150,000 in 1990 (see Fig. 4.4). It is only in Germany that coal industry employment exceeds 100,000 workers. The

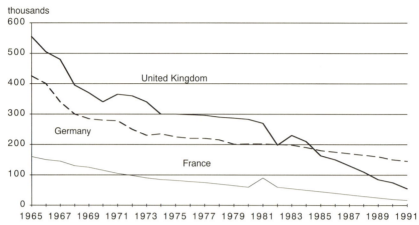

Fig. 4.4. Employment in the coal sector in France, Germany, and the UK

scale of the public support to the coal sector in Germany, with public support per employee considerably higher than average wages, should present ample room for compensatory arrangements for the coal communities. Still, it has until now been very difficult to gain political backing for measures that would significantly reduce production and public support to the German coal industry.

In Central Europe the scale of the employment issue is larger. The work-force currently stands at more than 350,000 workers, having contracted considerably since the late 1980s. The reduction in the number of coal miners has taken place without major strikes, but the relatively rapid mine closures have contributed to an apparent shift in political sentiments, leading to the more cautious approach to economic reforms in general and energy-sector reforms in particular. This is especially true for Poland. In the spring of 1994, for example, the government's proposal for a restructuring of part of the electricity sector provoked a strike amongst the workers in the lignite mines, which feed about 40 per cent of total electricity generation in the country. Proposals for reform of the electricity sector were withdrawn (IEA, 1995*a*).

The coal trade unions in Central Europe still represent an important political force. They are, as in Western Europe, well organized and they possess an important 'weapon' through the continued pivotal role of coal in the energy systems of Poland and the Czech and Slovak Republics. A miners' strike would bring these economies to a virtual standstill and could also have grave consequences in Hungary. In the short and medium term this reduces the political leeway for further large reductions in coal-industry employment, particularly since the large employment reductions that already have been made were generally accepted on the grounds of shrinking coal demand, environmental concerns, and obvious overstaffing and inefficiencies.

Security of Supply

Security of supply is, after employment considerations, the main argument put forward in favour of continued support to the coal industry. This argument too is gradually losing ground. West European production capacity has been dwindling and is currently only 10 per cent of total primary energy supplies. Therefore, indigenous coal no longer represents a buffer of any significance in case of a disruption of energy supplies from external sources, e.g. crude oil from the Middle East or natural gas from Russia. Moreover, the production capacities of coal-exporting countries like South Africa, Australia, Canada, Colombia, and other suppliers are probably elastic, i.e. additional supplies can be provided without upward pressure in the international price level. The infrastructure for the coal trade has over the past ten to fifteen years been greatly expanded, with European ports and handling facilities having the capacity to take considerable larger quantities of coal imports. It can also be argued, with reference to historical facts, that indigenous coal supplies are often disrupted by labour conflicts, and hence do not represent a more secure supply source than imports.

The security argument appears to be stronger in the case of the Central European countries. All have modest reserves of oil and natural gas, and their energy imports are to a large extent locked into existing pipeline systems from Russia and the former Yugoslavia. Moreover, the electricity systems of the Czech and Slovak Republics and Hungary depend on nuclear power of Soviet design (though currently with good performance with respect to regularity and safety), which also contributes to their dependence on foreign supplies and technical assistance. Security of supply in Central Europe is therefore a political priority on the same footing as improvement of economic efficiency in energy production and use. In addition, in these countries balance-of-payments considerations could favour maintaining indigenous production of energy.

Public Finances

A growing differential between domestic and international coal prices caused escalating economic losses in West European coal mining from 1986 to 1993, despite a 55 per cent improvement in labour productivity. The requirements for state aid therefore increased sharply. By 1992 PSE was about $10 bn for

Germany, United Kingdom, France, and Spain combined, of which a large part is now financed by public budgets (after the abolition of the 'Kohlepfennig' in Germany). In addition, some $10 bn is aid not benefiting current production, see Box 4.1. This compares to a total deficit in general government financial balances of $180 bn for the same countries in 1992 (OECD, 1994). Hence the support to the coal sector is significant even at the macroeconomic level. This is particularly the case for Germany, where coal support (PSE and aid not benefiting current production) was at a level equivalent to 20 per cent of the public financial deficit. The recent deterioration of the financial balance sheet, following the reunification with the new Länder in the former GDR, makes the coal sector a possible target for reduction in public support.

In Central Europe too, the status of public finances is a strong argument in favour of reducing the production capacity and financial aid to coal industries. Even if budget deficits have been reduced in recent years, there is still a fine balance between social stability, achieved through public spending, and macro-economic stability and economic reforms requiring prudence in governmental expenditures. Moreover, the financial requirements for investments in infrastructure and environmental clean-up are enormous. There will continue to be pressure to improve efficiency in the allocation of funds. No firm data exist on the scale of public support to the coal industry in Central Europe, but Radetzki (1994*b*) indicates that subsidies might have been $1 bn in Poland in 1992. Our rough estimate of subsidies in the Czech and Slovak Republics, points to a total subsidy in 1992 of some $250 m. These figures are small in comparison with support granted by West European governments, but they are still significant in terms of the size of the Central European economies and the precariousness of their public finances.

Electricity-sector Reforms

A major part of coal production in both Western and Central Europe is delivered to the electricity sector. Hence, any reforms in the electricity sector would have implications for the coal sector; conversely, coal policies may have bearings on electricity-sector reforms. Opening of the electricity markets to competition requires coal prices for electricity generation to be aligned to world-market levels, and flexible purchase obligations for domestically produced coal. Large industrial users of electricity in Germany have been pressing for policy changes in this direction, and the changes adopted open the way for a liberalization of the electricity sector. The privatization of the electricity industry in the UK was a major cause of the massive pit closures from 1992 to 1994.

These examples do not imply that introducing competition into the electricity sector will necessarily mean the end to most European coal production. However, if uneconomic coal mines are to survive under market conditions where electricity generators have full freedom of choice in investments and fuel purchase, the direct public financial support to the coal industry would have to be considerably increased, as has been decided in Germany. This means that the financing of the coal industry would no longer be shared between energy consumers and public budgets: the public budgets would have to bear it fully.

Environment

Coal has the image of being a 'dirty fuel' which causes environmental damage in production and consumption. In Western Europe, modernization of coal mining has significantly reduced emissions, effluents, and accidents at production sites. Air pollution from coal combustion is also increasingly being controlled by modern technology, but desulphurization devices add to the costs of coal use. As emission standards become more stringent, the competitive position of coal relative to other fuels deteriorates.

Environmental regulations have a bearing on total coal use, from domestic production and from imports. Depending on the quality of indigenous coal, emission limits may in particular affect national coal production. In the UK, for example, the requirements of the EU Large Combustion Plant Directive has contributed to the collapse of the British coal industry (Newbery, 1994), since indigenous coal has a relatively high sulphur content. In Germany, on the other hand, the strict national environmental regulations have no direct effect on coal production and use, because of protection of the coal and the electricity sectors. The costs of environmental measures in Germany are covered by electricity prices, due to the absence of competition in the electricity market. Since electricity generators are obliged to purchase minimum amounts of German-produced coal, for a major part of coal use in German electricity generation there is no direct competition between coal and other fuels.

In Central Europe, the extensive environmental consequences of coal mining (lignite and hard coal), both at production sites and in end-use, represent a critical factor for the future of the coal industry. The investment and other measures required in order to bring the environmental standards in production and the use of coal to acceptable levels are in many cases so costly that alternative energy supplies to coal will be preferred.

Finally, the challenges related to global warming and CO_2 may in the future have major consequences for coal use, imports, and production. Since there is no economically viable technological solution for restraining CO_2 emissions from combustion, the only answer is to impose measures that reduce the share of coal, and other fossil fuels, in the total energy balance.

Concluding Remarks

Coal production will continue to decline both in Western and in Central Europe. In Western Europe security of supply is no longer a weighty argument for keeping alive an industrial activity in which each employee is subsidized annually to the equivalent of US$ 60,000. The increasing problems of balancing public-sector revenues and expenses represent a major pressure for a continued reduction of subsidies, possibly an entire phase-out of coal production in Germany and France, with only a limited production to be maintained in UK and Spain. The time perspective for this phase out will depend primarily on the political weight given to employment and social problems that such a policy will create.

In Central Europe, and particularly in Poland, a larger part of the coal production is economically viable. Still, there are strong economic and environmental arguments for reducing production drastically. In the short and medium term, however, both employment and security-of-supply considerations may work against any swift changes, in addition to the substantial retrenchment that has already taken place.

5
Energy-Sector Developments: Natural Gas

Natural gas is a relatively new fuel in Europe, in the early 1960s accounting for less than 2 per cent of total energy use. The share grew rapidly from the late 1960s. Although growth slowed somewhat after the mid-1970s, natural gas has maintained its position as the most expansive part of the European energy market. Its increased use has reduced dependence on imported oil. Compared to other fossil fuels, it has certain environmental advantages which, in particular since the mid-1980s, have encouraged the increase in gas consumption.

In Europe as elsewhere, natural-gas use initially was based on local production. When local reserves were supplemented by exports from the Netherlands from the late 1960s, transport distances increased, but most of the gas trade still remained regional. The European gas industry became international with the advent of imports to the Continent from Norway, the former Soviet Union, and from Algeria. Since the late 1970s the growth in continental gas markets has been based on imports. Transport distances have increased as have investment requirements for bringing new gas to the market.

Due to the geographical concentration of the resource base in the UK, Norway, the Netherlands, Russia, and Algeria, gas markets have attracted substantial political attention, particularly in the gas-exporting countries. Both the Netherlands and Norway have government ownership of major gas companies. However, in most gas-consuming countries too, governments seek to influence overall gas strategies out of concern for security of supply, foreign policy, industrial and regional policies, and for environmental reasons.

Gas transmission and distribution have traditionally been seen as 'natural monopolies', requiring regulation and area protection, although there are wide differences among the European countries in the structure of the gas industry and the role of government regulation. The structure is now changing. Competition has been introduced in the UK following privatization; in Germany a new transmission company is challenging the position of the established companies; and there is pressure from the EU to open national markets for competition.

In Central Europe, gas from domestic production is relatively limited, and imports come only from Russia. Gas was formerly traded within the CMEA framework. As was the case with oil, a combination of Soviet export policies and foreign currency restrictions severely hampered the development of the national gas markets. In particular in Poland and the former Czechoslovakia, the emphasis was on utilizing the domestic resources of coal and lignite for electricity generation, in major industries, and in district heating systems for the residential sector. Today, in a more market-based economy, there is a potential for greater use of gas, substituting for solid fuels. Increased use of gas could make a major contribution towards reducing the environmental impact of energy use and improving overall energy efficiency.

5.1. Supply and Policies in Western Europe

Historical Background

The emphasis of government policies affecting the gas sector has changed over time. There have been four main periods in European gas-market policies (Stern, 1990).

1. *The Dutch period*: in the 1960s, international gas trade in Europe was dominated by the Netherlands. Gas markets developed rapidly, with emphasis on bulk users in electricity generation, in large-scale industry, and the conversion of town-gas networks to natural gas. Rapid development of the market was made possible by pricing gas competitively.

2. *The premium fuel*: in the early 1970s and until the mid-1980s North Sea reserves were developed in an atmosphere of increasing concern about long-term supply possibilities and the costs of new supplies. Existing long-term agreements were renegotiated. Gas prices increased and became linked to oil prices. Fears of scarcity led gas to be considered as a 'premium' fuel which should be reserved for high-value uses. Security of supply became an issue. This was first of all related to imports from the Soviet Union and the question of diversification of supplies received more attention. Higher gas prices and policies favouring nuclear power and coal reduced gas use in electricity generation, and growth in the total market for gas slowed down.

3. *Maturity and new supplies*: growth in gas consumption slowed down after the mid-1980s. At the same time, supplies became ample and oil and gas prices fell to a lower level than earlier expected. Reserve additions in the Netherlands

and in the North Sea boosted new supplies and earlier concerns about a possible scarcity of gas supplies waned. The issue of security of supply also received less attention. Several countries saw a revival of gas use in electricity generation due to technology improvements, lower gas prices, and environmental obstacles to nuclear power and coal use in electricity generation.

4. *New security-of-supply concerns*: in the 1990s, the European gas industry is facing new security-of-supply problems. The discussion of the 1980s concerned the potential political leverage of the former Soviet Union if the Western European countries became too dependent on imports from the East. In the 1990s, following the collapse of the Communist system, the disintegration of the Soviet Union and the increasingly unstable political situation in Algeria have given rise to new fears about the reliability of supplies from these sources. The fears are now that the governments in Russia or Algeria may lose control over exports or that civil war may interrupt exports.

Main Trends in Gas Consumption

Western Europe: From High to Lower Growth

The share of gas in total West European energy use increased from about 4 per cent in the late 1960s to 14 per cent in the mid-1970s. Since then, the increase has been far more moderate, with the share arriving at 18 per cent in 1994. From the late 1960s and into the early 1970s, the fastest growing part of the gas market was in industry and electricity generation. The expansion of the gas transmission systems presupposed bulk users of gas. In electricity generation, German and Dutch utilities switched from oil to gas use in existing power plants. Then, after the first oil-price increases in 1973/4, the 'premium fuel' philosophy, in combination with higher gas prices, resulted in a decline in gas use for electricity generation. From the mid-1970s, coal saw a renaissance, and the nuclear programmes also started to have significant impact on the fuel mix in electricity. Growth in gas markets became concentrated on the residential/commercial market, i.e. the small consumers. Fine-mesh gas distribution systems have been expanded, and gas has since the mid-1970s continued to gain market share in the residential/commercial market. There have also been continued gains in the share of gas in industrial energy use, but total industrial energy use has been on the decline because of higher efficiency and—not least—a restructuring with less emphasis on energy-intensive sectors.

From the late 1980s, gas use for electricity generation has seen a revival in a few countries, with the UK, Italy, and Belgium being the most prominent. In the UK, gas has become the preferred fuel for new electricity generation. For

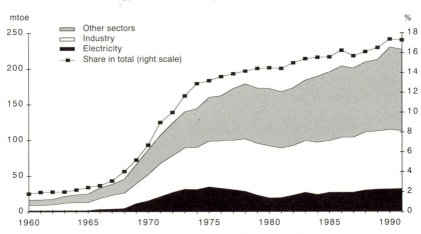

Fig. 5.1. Total West European gas demand and share in total energy use

the privatized electricity companies and other actors, gas is economically attractive compared to coal because of the capital requirements in combination with relatively low gas prices. It provides a faster payback on investments than coal, which for the private investor in the UK electricity sector constitutes an important criterion. In addition to its economic advantages, gas use in electricity generation in both Italy and Belgium is favoured because of the environmental advantages.

Central Europe: From Growth to Decline

The Visegrad countries have a share of gas in total energy use at almost the same level as in the West European countries, and the share has increased at about the same speed as in the West. However, these similarities conceal some important differences in the structure of gas use between Central and Western Europe.

In Central Europe, gas use is traditionally dominated by industry: industry accounted for about half of gas consumption, against only about one-third in Western Europe. Gas use in electricity generation varies considerably from country to country. In Poland, there is practically no such use, whereas both the Czech Republic and Hungary use gas in this sector. Even in these two countries, gas plays only a minor role. Much of their gas use in the electricity sector is for combined heat and power production in industry for the residential sector. In the centrally planned systems, individual gas heating had a low priority, as had individual heating with oil.

Since 1990, gas use in the Visegrad countries has declined sharply. How-

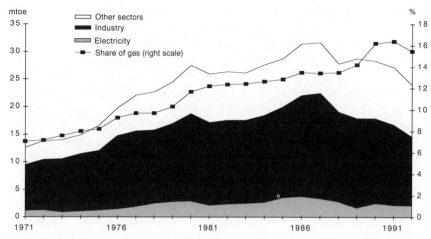

FIG. 5.2. Total Central European gas demand and share in total energy use
Source: IEA (1994*c*)

ever, this decline has been smaller than for total energy use, and the share of gas in total primary energy requirements continued to increase after 1990. The industries using gas have been somewhat less vulnerable to deep structural changes than the large coal consumers, such as the iron and steel industry. After the collapse of the centrally planned system, gas has found new markets both in industry and in the residential sector. In the medium to long term there is also in Central Europe a large potential market for gas use in electricity generation, not least because of the environmental advantages of gas compared to coal and lignite. Although the background is different, we may see in this area a parallel to the developments in the West European electricity sector.

Supplies

The West European gas industry started to develop in areas close to the centres of production. In the first phases, the majority of supplies came from on-shore fields in Germany, the Netherlands, Northern Italy, and Southern France. The UK market was developed after the discoveries of the gas fields in the Southern Basin close to shore. The discovery and development of the huge Groningen field in the Netherlands in the early 1960s made the European gas industry international, as Dutch gas was exported to Belgium, Germany, France, and even to Italy. The Netherlands played in the 1960s a key role in inter-European gas trade (Peebles, 1980). A further impetus to the market came with the development of the off-shore reserves in the North Sea and with long-distance transport of gas from Algeria and the former Soviet Union.

In the early 1990s, almost 70 per cent of total Western European gas supplies originated in four countries: the Netherlands, Norway, Russia, and Algeria, with most of the remainder coming from the UK, Italy, and Germany. The current major suppliers, Norway, Russia, and Algeria, all have the natural reserve potential to satisfy continued increases in gas demand and to compensate for the decline in production in other countries. The security implications of increasing imports from Russia or Algeria could act as a brake on substantial increases in contractual imports from these sources. Exports from the Netherlands will remain important over the next decade, but are unlikely to increase past their present levels. Once again, the question of the future composition of supplies, with a proper balance between economic and security-of-supply considerations, is on the agenda.

All of the major gas-consuming countries in Western Europe depend on imports of gas, although the UK is almost self-sufficient, while Germany and France are heavily dependent (see Fig. 5.3). Gas production in France is in decline and there are no prospects of finding new major reserves. In Germany, indigenous production has a relatively low share in total supplies and production is likely to decline in the long term. Import dependence is an important factor for the major gas-consuming countries. It is the general perception of the major gas companies and governments in the continental countries that security of supply and the possibilities of obtaining supplies at reasonable cost require the concentration of market control on a few players.

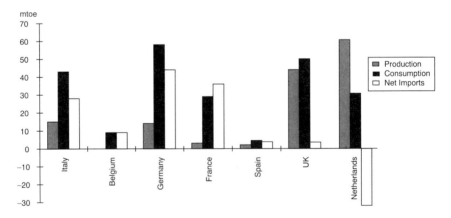

Fɪɢ. 5.3. Production, consumption, and net imports of gas, major West European countries, 1992

In some countries, dependence on imports has resulted in governments taking an active role in overall gas strategies, as in Belgium, France, and Italy, where the importing companies are state-owned or under government influence. In Germany, imports are undertaken by a few major gas companies, which all are private, and there is no direct government control of import policies. The strength of the major gas companies in this market is, however, partly due to German legislation, which allows the established companies to share the downstream market among themselves geographically. In the UK, the government has intervened in projects for imports of gas. In the British case, government intervention is not guided by security-of-supply considerations to the same extent as on the Continent, but mainly by concern about the production and activity level in the UK off-shore sector and by balance-of-payments considerations.

In Central European countries, governments are likely to retain both significant ownership of and influence on the major gas companies. The emphasis has been on a restructuring of the gas industry, breaking up the former state enterprises into separate production, transmission, and distribution companies. In Poland, the restructuring of the gas sector is not yet accomplished, but it is likely that the government will remain the majority shareholder of a future national gas import and transmission company. The policy of the Czech Republic is for a certain time to retain the single national supply and transmission company, Transgas, as totally state-owned. If Transgas at a later stage is included in the privatization programme, for example through the voucher system offering ownership to the general public, it is current policy to maintain

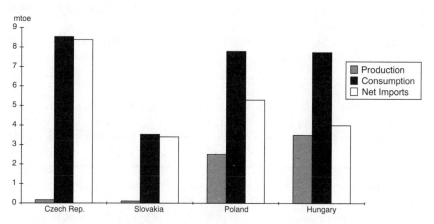

FIG. 5.4. Production, consumption, and net imports of gas, Central Europe, 1992

a state majority interest in the company. The reason is the company's strategic importance in supplying gas to the Czech market and in transporting Russian gas to West European markets.

Successive Hungarian governments have had a policy of maintaining strategic shareholdings in the national transmission company and in the regional gas distribution companies as well. This policy was modified somewhat in late 1994, when the government embarked on a privatization programme for the gas distribution companies. The municipalities have retained about 50 per cent of the shares, the remainder being sold off to Western energy companies. The Hungarian state will, however, retain the right to control strategic company decisions such as mergers and acquisitions by holding a 'golden share' in the companies. In Hungary's gas transmission company—which is the integrated former national oil and gas company—the state will retain a minority share sufficient to give it a blocking minority right in important decisions affecting the company and overall gas policies.

Gas was introduced in the former Czechoslovakia in conjunction with the construction of the Soviet pipeline system intended for exports to the West. The pipelines transit both the Czech and Slovak Republics and supply them with practically all of their gas needs. In Poland and Hungary, gas use was originally based on their own production, later supplemented by imports from Russia. In all four countries, gas imports were originally part of the bilateral trading system between Eastern European countries and the Soviet Union, but prices and payment conditions for imports have since been changed to resemble those in the West. There are, however, special arrangements in the Czech Republic and in Slovakia: both countries receive a major part of their gas needs from Russia as payment for the maintenance and operation of the transit pipelines to the West.

In the following, a brief overview is provided of the gas export policies of the major exporting countries.

The Netherlands

Because of the economic importance of gas, the Dutch Government has maintained strong influence over its gas sector. The export strategy is determined jointly by the gas industry and the government. A determinant factor for the Netherlands' export policies are the long-term needs of the domestic market. Gas exports have been allowed only to the extent that they have been compatible with domestic requirements. There is in the Netherlands a depletion policy, according to which total demand for gas is evaluated against the reserves. Export policies and gas policies for the domestic market are also influenced by the need for government revenues and for export revenues.

Export policies have consequently changed direction several times. Some of these changes have been attributed to revised estimates of remaining gas reserves, but there is undoubtedly often a political reasoning behind changes in the published gas reserve estimates.

We may distinguish three phases in Dutch gas policies. The first phase saw rapid expansion related to the build-up of the domestic market and of exports in the 1960s. The second, in the 1970s, was a period where conservation of gas reserves was at the forefront. Typical of this period was the firm ceiling imposed on exports and limitations on the use of gas for electricity generation. In the third phase, from the mid-1980s, the gas reserve conservation argument lost in significance, and public revenue and balance-of-payments considerations regained importance.

In the 1970s, gas was seen as a premium fuel whose use in electricity generation and other 'low' value sectors should be restricted. An important objective was to secure sufficient gas supplies for the domestic market and to maintain the strategic role of the Groningen field. It was decided to commit no further export volumes and to cease exports towards the end of the century. Domestic supplies were even supplemented with imports from Norway.

Originally, gas was exported under long-term contracts at almost fixed prices. A complete turn-about in export-pricing policies took place in the late 1970s, triggered by a political crisis. Dutch consumers complained that they were paying higher prices than foreigners for gas from the Netherlands. Furthermore, the second oil-price shock in the late 1970s widened the gap between Dutch export prices and oil prices. The government threatened to cut off exports, appointed a gas export negotiator, and came to play a leading role in the renegotiations which succeeded in obtaining higher prices. A breakthrough was the linking of gas prices to oil prices.

In the 1980s, demand for Dutch gas was lower than expected, leading to a policy revision. New gas volumes were made available for power generation and existing export contracts were extended. From the mid-1970s, domestic gas consumption had started to decline because of less use of gas in power generation, higher energy efficiency, and changes in the structure of industry. In the same period, the official reserve estimates were revised upwards—a process which has continued since. The improved reserve situation was the justification for new volumes of gas being made available for exports, but official Dutch policy is still to cease all exports when the existing export contracts expire in the years around 2010. An increase in annual volumes for export is not to be expected, but it is possible—as demonstrated in the past—that export contracts will be extended in order to maintain export revenues. It is likely that in the future, as in the past, there will be significant gas reserve additions in the Netherlands (Stern, 1990).

Norway

Statoil, the state-owned oil and gas company, has a central position in all commercial aspects of gas exports and in implementing Norway's overall gas export strategies, which are determined with the government. It participates in the most important oil- and gas-field developments. All gas exports are contracted centrally through a gas negotiation committee, the 'GFU', in which Statoil has a leading position. The GFU decides on the allocation of production from the fields and also centralizes commercial negotiations for the gas sales. The government has through the licensing policy an important role in the strategy formulation for the gas sector and this influence is cemented through the central role of Statoil (Estrada *et al.*, 1988; Stern, 1990).

In contrast to the Netherlands, the main Norwegian concern is not related to the depletion of the resource base, but rather to maintaining the balance between the development of oil versus gas and the effects of off-shore activity on Norwegian industry, employment, government revenues, and balance of payments. Norwegian gas export policies have developed through different phases. In the 1980s it was signalled that gas prices should be high, otherwise efforts would be concentrated on developing oil reserves instead of gas. This policy message was contradicted by the results of exploration, which showed that gas was the predominant resource. Thus, Norway had to reverse its earlier policies and increase the importance attached to gas exports. This was underlined by the discovery of the huge Troll field, the largest single West European gas field discovered since the Dutch Groningen field.

In the first half of the 1980s, the Troll development was seen as being of strategic importance in resolving the security of supply problem for continental Europe. Both French and German buyers had a commercial need for new supplies from other sources than the USSR or Algeria. The production from another important field, Sleipner, was intended to be sold to the UK market. A contract was in fact negotiated for sales of gas from Sleipner to British Gas, but the contract met political obstacles. Most of the contract details had been finalized when the Thatcher Government in 1985 blocked new imports from Norway. Apparently, the motivation behind this interference in commercial negotiations was to give priority to the development of UK's own off-shore gas resources. The Sleipner field was sold instead to continental buyers as part of the complex of gas export contracts linked to the development of the Troll field.

The size of the Troll field offers the Norwegian producers greater freedom to phase in smaller fields, using the Troll reserves as a 'buffer'. Instead of selling gas on a field-by-field basis, gas can be marketed from a variety of fields using Troll as a volume guarantee. This has given the field an important commercial role. When the Troll contracts were concluded in 1986, oil prices had fallen sharply, demonstrating the huge financial risks involved in this project. It was probably only because of Norwegian Government involvement (the government held directly or indirectly through Statoil a 75 per cent share in the field) that it was possible for the project to go ahead. Private investors would have been less willing to assume the long-term risks involved.

Since the early 1990s, Norwegian gas exporters have taken a more active role in the downstream marketing of gas. Statoil has, for example, acquired a minority interest in the transmission company of former East Germany, VNG, and has also established a gas marketing company, together with Norsk Hydro, aimed at the UK market.

Ex-USSR

The former Soviet Union possesses huge gas reserves which have been developed to supply the energy needs of the country itself, its former allies in Eastern and Central Europe, and also to provide much-needed foreign currency earnings. Imports of Soviet gas, in particular to West Germany, triggered in the early 1980s a debate between the United States and European Governments about security-of-supply problems related to heavy reliance on USSR. The row over the Soviet imports was more far-reaching than concerns for the West European gas markets: it became a cold war issue debated not only in the EC and the IEA, but also in NATO.

The extensions of the Dutch export contracts, and in particular the Troll discovery and eventual sale to continental buyers, imply that the 'overdependence' on Soviet gas imports did not materialize. The commercial arrangements reached between the Soviet and the West European importers also invalidated fears that the Soviet exporter would undermine price levels in the West. Soviet exports were priced close to the average gas import prices in West Europe.

Since the collapse of the Soviet regime, Western concerns about gas supplies from Russia have shifted. Although there is an important need for reinvestment in the gas sector, as elsewhere in the Russian economy, there are no serious concerns about the technical capability of the Russian system to produce and transport gas to customers in the West. Domestic consumption of gas in Russia and in other parts of the former Soviet Union has declined because of the severe industrial recession, in particular reductions in production in energy-intensive industries. This has more than offset the deterioration in production capacity. In the longer term, the Russian resource base could sustain significant increases in production and exports, although there are some doubts about costs. In the short to medium term there are two sets of worrying factors for the West. One is again the question of the security of supplies from Russia. This time, security concerns are not linked to fears of becoming dependent on a major international political power, but to the possibility of supply disruptions triggered by the unstable political situation in Russia itself and in the republics transited by the pipelines supplying the West. The most worrying question is the transit of gas through the Ukraine, where conflict over non-payment for Russian deliveries to the republic has occasionally threatened the transit of gas to the West. The fragility of the Ukraine's economy in general, and its energy economy in particular, carries the risk of serious long-term supply disruptions to West and some Central European countries.

The other question often posed concerns possible changes in Russian gas export strategy. There have been fears that declining oil exports and the insatiable currency needs would push the Russians to underbid the prevailing import price level in Western Europe in order to gain volumes and market shares. The gas sector has been reorganized, with gas exports remaining in the hands of one single company, Gazprom. The company is actively seeking new markets in the West and is building new alliances. A main purpose seems to be to gain more direct access to the end-users of gas, but so far without jeopardizing the pricing of the gas sold to the established gas companies in the importing countries. Gazprom has made alliances with other gas companies, for example, in Finland, in the Baltic states, with an important industrial company in Italy, and is also part of the consortia behind the construction of the first direct connection between UK and the Continent, the Interconnector linking UK to Belgium. Undoubtedly the most important step in this direction taken by Gazprom is the investor interest it has taken in a new German transmission company, Wingas. Wingas sells primarily Russian gas in competition with the established German gas companies. This strategy has created some 'gas-to-gas' competition in the German market, where formerly all of the companies operating in Germany respected the demarcation areas (Stern, 1995).

Algeria

Algeria is a densely populated country facing serious development and increasingly difficult political problems. Since the late 1980s these have threatened to develop into a full-scale civil war that ultimately could lead to an overthrow of the military regime. Oil and gas export earnings are essential to pay for import needs, but the political instability has made other countries reluctant to rely on gas imports from Algeria. Consequently Algeria has obtained lower export volumes than its production potential would indicate (Stern, 1984).

On the commercial side, the Algerian pricing policy has been a legendary cause for conflict. All of its long-term export contracts—in contrast to short-term spot sales—have been the subject of protracted disputes, renegotiations, and arbitration. In the late 1970s, the Algerians advocated a 'hard-line' pricing policy, aimed at pricing gas at the Algerian border at a level close to crude oil parity. As a major part of the gas was exported as liquefied natural gas (LNG) with subsequent transport and regasification costs, this policy resulted in a substantial higher price for gas than oil at the burner tip. In addition to its substantial LNG export capacity of about 30 BCM per year, a pipeline (Transmed I) was built to Italy.

Although the principle of parity between the export price for gas and crude oil was abandoned, Algeria pursued a 'high-price' policy during most of the 1980s. This resulted in protracted conflicts with customers—the USA, France, Italy, Belgium, and Spain. In some instances, these conflicts were solved by government intervention in the importing countries. In Italy and France, negotiations between Algeria and the importing gas companies, SNAM and Gaz de France respectively, stalled first of all on the question of pricing. 'Political' solutions were found to these conflicts when the governments intervened and finalized the gas import contracts at price conditions out of line with the market. Both governments were motivated by concerns for their industrial interests in Algeria, and in the case of France also by a more general wish not to jeopardize the complex bilateral political relationship with its former colony.

Algeria has maintained a hard-line policy in pricing, but there has been some softening of its stance. Contracts on commercial conditions were negotiated with the USA and new volumes contracted, in particular for the Italian and Spanish markets. The pipeline connection with Italy has been strengthened and a new major pipeline project linking Algeria through Morocco to Spain is under development. This connection, the Maghreb Pipeline, will in the first instance serve Morocco and Spain, but the pipeline could in a longer time perspective supply Algerian pipeline gas to other parts of Europe, transiting through France. There is no doubt that urgent internal economic problems have contributed to a certain shift in Algeria's gas export policy. Balance-of-payments considerations, in combination with the relatively low oil and gas prices after 1986, have led the government to seek short-term volume gains at the expense of the previous almost prohibitive pricing policy.

The political clouds over the Algerian regime and the resulting uncertainty about the sustainability of gas exports are different from those in Russia.

Following the halt to the democratization process and cancellation of the results of general elections in 1992, which had given a majority to fundamentalist Muslim opposition movements, the country has moved into a situation resembling civil war. The military government is gradually losing control of major parts of the country and there is ultimately a threat that oil and gas installations may be sabotaged. For France, Italy, and Spain, who all—to varying degrees—depend on gas imports from Algeria, this situation constitutes a real security-of-supply problem.

5.2. Market Structure and Policy in Gas-Consuming Countries

Western Europe

The organization of the European gas industry varies significantly between countries. With the exception of the UK, there is an element of public ownership in all countries. In distribution of gas, local authorities own or have the controlling interests in the undertakings in most countries. For imports and transmission, central governments have in some countries the full or a partial ownership, but private interests are more widespread in this layer than in distribution, see Table 5.1 for a simplified presentation.

With the exception of the UK and Germany, there is only one company in each country responsible for imports and transmission of gas. Major international oil companies have, apart from their production interests, owner interests in some of the companies responsible for imports of gas. Oil-company

Table 5.1. *Ownership structure of West European gas industries*

100% private	Mixed	100% publicly owned
Transmission		
Austria	Spain	Italy
Belgium	France	Denmark
Germany	Netherlands	
(major importers)		
UK		
Distribution		
UK	Italy	France
	Germany	Austria
	Spain	Denmark
	Belgium	Netherlands

downstream interests in the gas industry are strong in Germany, Belgium, the Netherlands, and also in the UK for the new gas supply companies competing with British Gas. Shell, Esso, Mobil, Texaco, and BP all have shares in Ruhrgas, the dominant gas company in Germany. Shell and Exxon own together 50 per cent of the Dutch Gasunie, and Shell has also a share in the Belgian importing and transmission company, Distrigaz.

In all these countries, with the exception of the UK, gas companies enjoy geographical protection of their supply areas. Local distribution is considered a 'natural monopoly', and as mentioned, there is usually only one importing and transmission company. In Germany, where several companies are importers and own transmission pipelines, the established transmission companies oper-ate under agreements which provide area protection and effectively divide the country into separate spheres of business interest.

Monopolies in transmission and distribution, together with concentration of production on a few major suppliers, have meant that gas-to-gas competition is practically non-existent, again with the exception of the UK market. Policy interest has, however, increasingly focused on the possibilities for changing the institutional framework so as to create more competition. In the UK, following the privatization of British Gas, the introduction of TPA and removal of the monopoly nights of British Gas, new gas-marketing companies have been established with the participation of gas-producing companies. The Russian Gazprom's owner share in the German company Wingas is a new phenomenon which is weakening the traditional area protection of the estab-lished companies. An overview of the structure and regulation of the gas industry in the major gas consuming countries is given in Table 5.2.

Despite some common features, there is considerable variation in the struc-ture and regulation of the gas industries in individual countries. The situation in some of the major gas-consuming countries is discussed in more detail in the following sections.

The Netherlands

The Netherlands has a tradition of close interaction between the government and the gas industry. The state owns 50 per cent of the national transmission company, Gasunie, and decides with the company the overall strategy for gas production, management of reserves, and exports. Gas production is under-taken by private companies. The largest gas producer in the country is NAM, which owns the Groningen field, the single largest field in the country. NAM is owned by Shell and Esso. Furthermore, NAM holds a 50 per cent share in Gasunie.

Gasunie is responsible for gas purchasing from Dutch fields and owns the

Table 5.2. *Organization and regulation in major gas-consuming countries*

	Organization	Government regulation and policy
Germany	Several layers, a small number of big merchants and transmission companies, many distribution companies. Some vertical integration by owner shares	There is no direct government control. Gas transmission companies have the right—as an exception to the general monopoly legislation—to agree on exclusive areas: the 'demarcation' agreement. Distribution companies have area protection as supply monopolies. Prices to small consumers are publicly controlled (cost-plus basis)
France	One dominant vertically integrated company, some private interests in limited parts of the country	Monopoly protected by law, framework for general policies, and coordination with the government
UK	Originally British Gas covered the entire country as a vertically integrated company. New entrants build own pipelines or use British Gas's system	After privatization of British Gas no area protection, TPA, regulatory pressure for more competition. Pricing policies of companies with a dominant market position—like British Gas—are controlled by a regulatory body
The Netherlands	Two layers of companies with one merchant and transmission company and several distribution companies	Strong government influence on overall gas strategies, area protection of distribution
Italy	Two layers of companies with one merchant and transmission company and many distribution companies	Loose framework for overall strategy, no monopoly protection for transmission but for distribution

transmission system. All gas from the Groningen field is sold to Gasunie, as is most other gas. The company has by law the right to receive a first offer of all Dutch gas and has a *de facto* but not legal monopoly on transmission. Some foreign gas companies buy Dutch gas and obtain transportation rights through Gasunie's grid.

Gasunie has the exclusive right to supply domestic distribution companies,

Table 5.3. *Organizational structure of the gas industry in the Netherlands*

Activity	Company structure	Ownership
Production	International and national oil companies	100% private
Imports, exports, wholesale, and transmission	One company: Gasunie	50% state, 50% private
Distribution	About 35 companies	100% owned by local public authorities

which in turn have monopoly rights to supply small- and medium-sized users. Industrial users of gas are generally supplied directly by Gasunie. Pricing policies have occasionally been used to favour specific sectors or to support the energy policy or other aims. However, special tariffs for the use of gas in greenhouses have come under attack from the EU Commission for containing a subsidy element. There are also special tariffs for the sale of gas to small-scale co-generation plants, as a means of supporting the government's general energy and environmental policies.

The organizational structure of the Dutch gas industry has been in place since it began in the late 1960s. Apart from a horizontal concentration of distribution, the structure has remained stable. Distribution companies have endeavoured to obtain more freedom in tariff setting and also increased their negotiating leverage against Gasunie. There appears, however, to exist clear political support to maintain the central role of Gasunie and to use Gasunie directly and indirectly to implement government policies in the gas sector. In early 1996, the government outlined a plan for a liberalization of the gas industry. According to this, Gasunie's grid would be opened to TPA and large gas users, and distribution companies would have the right to purchase gas directly from Dutch producers or to import.

The overall strategic formulation of gas policies is centred on the 'Gas Marketing Plan', prepared each year by Gasunie in collaboration with the government. The Gas Marketing Plans are the basic instrument for signalling to the outside world the main features of the Dutch gas reserve situation, infrastructure investments, pricing policies, and not least the balance between sales to the domestic market and exports.

Germany

Unlike other West European countries with only one national transmission company, Germany has several. More than twenty companies own and operate

Table 5.4. *Organizational structure of the gas industry in Germany*

Activity	Company structure	Ownership
Production	International and national oil companies	100% private
Imports, wholesale, and inter-regional transmission	6 dominant companies	100% private
Wholesale and regional transmission	About 15 companies	Mainly private and mixed private/public
Distribution	50 regional and about 480 local companies	Mainly public and mixed private/public

high-pressure transmission grids, although only a small number have a central role in the supplies of Germany. The major importing companies are BEB, EWE, Thyssengas, Ruhrgas, VNG, and Wingas. Ruhrgas, the dominant transmission company and importer, has significant owner interests in half of the other transmission companies as well as in the international transit lines. Ruhrgas accounts either directly or through its affiliates for about 70 per cent of total German gas sales. The six big importers also hold controlling interests in the major part of the transmission system. They work as inter-regional companies, supplying a mix of regional transmission companies, local distribution companies, and towns, as well as major industrial customers (Schiffer, 1994).

Private ownership is dominant in the upper layers of the system, and there is increasing public ownership at the lower levels (see Table 5.1). The owners of the gas companies in imports and transmission are international oil companies, other German energy companies as electricity generators and coal producers, and some important German industrial conglomerates. Some public ownership is found in wholesale and regional transmission, which consists of transmission companies without direct access to imports. Some of these are controlled by local distribution companies, municipalities, and Länder interests. Others have private owners or a mix of private and public ownership. Local distribution companies, of which there are about 500, are mainly controlled by the municipalities. These companies are also to a varying extent owned by regional transmission and wholesale companies.

With the reorganization of the former East German energy industry, the transmission company, VNG, was privatized. Ruhrgas obtained a 35 per cent share of VNG. Through this and by owner shares in the local distribution companies, also privatized, Ruhrgas consolidated its position in the new

Länder. The need for investment in upgrading and expansion of the gas system has been used as an argument to give the major West German companies a leading role in the privatization process. Non-German energy companies have not managed to gain any substantial hold in the market, although foreigners obtained minority shares both in VNG and in some of the distribution companies. British Gas, ELF, Statoil, and Gazprom each obtained a 5 per cent share of VNG.

The Federal government has no owner interests in the gas sector, and it is difficult to define an official gas policy for Germany. The sector operates within the framework of the monopoly legislation and the licence system administered by municipalities. Germany's monopoly legislation allows gas and electricity companies to enter into geographical demarcation agreements. Such agreements have existed between the inter-regional companies from the very start of the German gas market, and effectively exclude competition between them. Regional transmission companies may have several suppliers, depending on the demarcation lines, but unless they border on the Netherlands, they can import only with the consent of one of the pipeline owners—a consent not normally granted. It has proved practically impossible to contest refusals of request for transport in the German courts on the grounds of abuse of dominant market position. At the local level, the right to distribute gas to small users requires a licence from the municipalities.

The right to construct transmission pipelines may be granted to any company under German energy law, on condition that the need for the line is proved. This right has been used by a new player in the German market, Wingas, which has been selling gas across demarcation lines both in the former East Germany and in the West. This company is owned in equal shares by the Russian gas export company, Gazprom, and the German energy and chemical group Wintershall. It supplies mainly Russian gas, some volumes of German gas produced by the Wintershall group, and has also obtained supplies from the UK to be delivered when the Interconnector is ready for use in 1998. Wingas is the only transmission company in Germany which has not signed any demarcation contracts with the other gas companies.

The Minister of Finance, who also is responsible for energy, proposed in 1993 changes to the existing legislation intended to open the German gas and electricity markets to more competition. These proposals met with strong opposition from the gas industry—as was the case also in the electricity industry—and were eventually abandoned. They were, however, revived in 1996 in a somewhat modified form in connection with the negotiations in the EU on the directive for liberalization of the electricity markets. It was proposed to abolish the demarcation agreements by a change of the monopoly legislation and weaken the area protection contained in the concession agreements for distribution companies. Originally, the facilitation of third-party access to the grid was also included. This most controversial part was left out when the proposals were relaunched in 1996. Parallel to the political process, the German Cartel Office has added to pressure on the regulatory system by trying to apply EU monopoly rules to attack parts of the demarcations agreements and also the concession agreements. None of the cases brought forward by the Cartel Office has so far been finally decided. An obstacle to establishing a more competitive market lies, however, in the ownership structure. The owner interests of inter-regional companies, and in particular those of Ruhrgas, in some of the regional transmission companies could impede any major changes.

France

The French natural gas system is highly concentrated and closely connected to the government. The fully state-owned Gaz de France (GdF) was created in 1946 as one of the elements of the reconstruction programme in the aftermath of the Second World War. It was the objective of the government and its Commisariat de Plan to keep close control of the economy by the nationalization of key financial and industrial sectors as well as energy. GdF has therefore the same political background as EdF for electricity and CdF for coal. Both EdF and GdF operate in close collaboration with the government and the central administration. A close-knit network of political and personal ties ensures that the general direction of GdF's policy, in regard to gas purchasing and investment and marketing, is in line with the policy of the government.

GdF as a gas importer has been and still is to a certain extent obliged to obey French foreign policy requirements. This is particularly clear in regard to the gas import arrangements with Algeria, where the French Government has intervened and forced GdF to accept price conditions outside the normally accepted commercial range. The government has also tried, but without any noticeable success, to link GdF's gas imports from the earlier Soviet Union into an economic package arrangement aimed at promoting French industrial exports.

GdF enjoys a legal monopoly of imports of gas and a monopoly of transmission, except for the central and southern parts of the country, where two transmission companies operate. Both of these, the SNGSO and CFM, are owned by GdF together with the two French oil companies, Elf and TOTAL. The two oil companies were formerly state owned, but have now been privatized. The complex arrangements for the central and southern parts of the

Table 5.5. *Organizational structure of the gas industry in France*

Activity	Company structure	Ownership
Production	National oil companies	Mixed private/state
Imports and transmission	One dominant company: Gaz de France (GdF)	100% state
	Two regional transmission companies	Mixed private/state
Distribution	22 municipal companies plus GdF	Mainly state

country were made to secure the French production companies' interests. GdF controls distribution nation-wide, with the exception of a relatively small number of towns which obtained the right to retain their local distribution companies when GdF was created.

Since 1973/4, French energy policies have been directed at reducing the dependence on imported energy. With high-cost domestic coal resources and a declining contribution from domestic oil and natural gas production, the nuclear programme was given priority. Not only was nuclear power to substitute for fossil-fuel use in power generation, but electricity should obtain a higher share of the end-user market at the expense of imported fuels. This policy implied that the government imposed a 'defensive' marketing and investment policy on GdF, in particular in the space heating market, compared to the much more active policy of EdF in the 1970s and early 1980s. In new houses, heating is usually by electricity.

The import monopoly of GdF is currently under attack by the EU Commission. This and other issues in relation to the European debate on TPA have been examined in the Mandil Report, elaborated by the ministry responsible for energy. The policy recommendations are similar for natural gas (i.e. GdF) and electricity (EdF). The Mandil Report (1993) suggests that major industrial energy users should be allowed to import directly for their own needs and to obtain transport rights in GdF's system on a negotiated basis. TPA is rejected as jeopardizing efficiency and security of supply. Major industrial gas consumers, who would use the right to import directly, would be obliged to negotiate the terms for the imported gas with GdF. Therefore, according to this cautious liberalization, GdF would retain its pivotal role in French gas supplies. The recommendations of the Mandil Report have been copied into the EU directive for the liberalization of the electricity sector, allowing for a 'single buyer' as an alternative to TPA.

Italy

The energy sector in Italy has a strong state dominance dating from nationalization under the Mussolini regime. The ENI group, a state holding company,
owns the national oil company AGIP and the gas transmission and merchant
company SNAM. The ENI group also owns a number of key industries, for
example in chemicals and in iron and steel production.

ENI has traditionally had a strong position in the Italian political system.
The state holding company may be described as 'a state within the state'. It has
managed to maintain some degree of independence from the influence of the
often rapidly changing governments, while at the same time being strongly
integrated with the Italian political establishment. One could say that the
strength of ENI is largely due to the often confused parliamentary situation.
The result has been the integration of the gas sector in the political system and
of gas policies in overall government policy. This has been the case for the
development programmes in the southern part of Italy. The government
decided in the early 1980s on a 'methanization' programme for these regions.
Extensions of the gas grid were promoted by government investment grants
and tax advantages and SNAM agreed to sell gas to distribution companies in
the South on more advantageous conditions than in the North. Major parts of
ENI are now being privatized and the role of the holding group has furthermore been weakened by parts of the top management's involvement in corrup-
tion. So far there are no plans to privatize SNAM.

SNAM owns the high-pressure transmission network, negotiates most of the
imports to Italy, and purchases all gas produced domestically. A major supplier
of gas to SNAM is AGIP, which holds a monopoly on gas production on-shore
in Northern Italy. SNAM has no legal monopoly on imports, and there is only
limited third-party access to the transmission system. Access has been granted
to ENEL, the state-owned electricity utility, which imports gas for its own use
in electricity generation. Large industrial users may also obtain the right to
transport. Of the numerous local distribution companies which purchase all
their gas from SNAM, 24 per cent are private, 34 per cent are owned by the
municipalities, and 32 per cent are part of the Italgas group. Italgas is a private
company in which SNAM has a decisive owner share.

The structure of the Italian gas industry (see Table 5.6) resembles that in the
Netherlands. *De facto* vertical integration is, however, stronger. SNAM has
owner interests in the largest otherwise privately owned group of distribution
companies, Italgas. Energy taxation policies have supported the gas sector in
Italy to a much larger extent than elsewhere in Europe, with the exception of
Denmark. Taxes on oil products have since the early 1970s been relatively

Table 5.6. *Organizational structure of the gas industry in Italy*

Activity	Company structure	Ownership
Production	State-owned AGIP for on-shore production, mixed for off-shore	Mainly state-owned (ENI)
Imports, wholesale, and transmission	One company: SNAM	100% state-owned (ENI)
Distribution	About 700 companies	Mixed private and state-owned

high, mainly for fiscal reasons. Of importance for the gas sector is the high tax on light heating oil. The much lower tax on natural gas gives gas a competitive edge in the residential market.

United Kingdom

UK gas policies fall into two distinct periods; before and after the privatization of British Gas in 1986 when the 100 per cent state-owned company was sold off to the public. British Gas was established after the Second World War as a vertically integrated company with a monopoly to purchase gas from the UK continental shelf and the sole right to import gas. On-shore the company held *de facto* monopoly rights to transmission and distribution for the entire country. Formal regulation of the gas sector was light-handed. There was only an informal control of pricing policies, which for the residential market were based on a cost-plus philosophy.

The first off-shore gas fields, the Southern Basin fields, came on-stream in the early 1970s. Gas producer prices were related to the costs of exploration and development. The bargaining position of British Gas as the sole buyer for a large market was underpinned by the requirement that gas produced in the UK North Sea should be landed in the UK. This requirement survived criticisms from the EC Commission and from producers.

For the development of off-shore gas supplies and transport systems, the British Government has in general terms followed a non-interventionist policy, apart from the potential threat to commercial freedom inherent in the landing requirement. It has, however, intervened in plans for imports of new supplies. In 1985 for example, the Thatcher Government blocked an agreement between British Gas and Norwegian suppliers for imports of gas from the Sleipner field. On the Norwegian side, the strategy was to sell the Sleipner gas to the UK and to supply the continental countries with new gas from the Troll structure.

British Gas, which at that time was still state-owned, saw a need to supplement indigenous production with new imports from Norway, *inter alia* because gas imports from the Norwegian part of Frigg were expected to decline substantially before the end of the 1990s. It thus came as a surprise that the Conservative Government had decided that British Gas did not need any new imports. The government's motivation was to secure a market for the development of UK gas reserves in the second half of the 1990s. Balance-of-payments considerations also played a role in the decision.

Public regulation has changed radically following the privatization of British Gas, which was sold off as a vertically integrated and highly profitable company. Because of the monopoly position of the now private company, it was decided in the Gas Act of 1986 to create a regulatory authority (OFGAS) to monitor British Gas's pricing policies. The Gas Act also stipulated the conditions for use by third parties of British Gas's grid for transportation. Third-party access to the grid existed before privatization but had never been effective. The policy aimed to open it up for competition from new entrants in the gas market, which in combination with public regulatory powers was considered as a justification for maintaining the vertical structure of British Gas.

Since then, there have been protracted conflicts between British Gas, gas consumers, OFGAS, and the Monopoly and Mergers Commission (MMC, the competition watchdog) on pricing policies for small- and medium-sized customers, on transport tariffs, and on the role of British Gas as a gas purchaser from the UK North Sea. MMC recommended that competition should be promoted by restricting British Gas from buying more than 90 per cent of the UK gas offered for sale. The government accepted this recommendation, although only for two years, but British Gas has committed itself to the 90 per cent rule, which was intended to provide a basis for gas-to-gas competition. OFGAS has continued to put pressure on British Gas, and the company has relinquished gas from existing contracts to other gas companies which have started operating on the UK market.

Competition from independent gas companies, of which some are owned by gas producers, has resulted in British Gas losing market shares. By the end of 1993, 30 per cent of the medium-sized user market was supplied by independent gas companies, against none only three years before. Altogether 78 per cent of the large users were supplied by independent companies, against only 9 per cent at the end of 1990. The loss of market share and increased price competition has put British Gas in a difficult situation. The company has an overhang of earlier gas purchase contracts signed with take-or-pay conditions and at high prices compared to the prevailing market conditions. This situa-

Table 5.7. *Organizational structure of the gas industry in the UK*

Activity	Company structure	Ownership
Production	International and national oil companies	100% private
Imports, wholesale, and transmission	British Gas and independent companies	100% private
Distribution	British Gas	100% private

tion, which threatens the financial position of the company, has parallels to what developed in the USA following deregulation of the gas market in the 1980s.

The household market will be opened gradually, with 100 per cent of the market to be made accessible to independent companies by 1998. Instead of breaking up the vertical structure of British Gas itself, independent gas companies will obtain the right to use British Gas's distribution systems and storage services necessary to serve small users of gas.

Central Europe

In Poland, the gas network was developed to receive and distribute imported gas from the Soviet Union and domestically produced gas from the south-eastern part of the country. The network consists of three separate systems: one for high calorific gas, most of which is imported from Russia, one for low calorific gas, and one for coke oven gas. The three networks cover most of Poland, but capacity is limited. Gas use is concentrated in industry, but in recent years residential gas use has been growing. There is a potential for gas use in electricity generation, in particular in small scale co-generation which currently is using coal (IEA, 1995*a*).

The public utility, POGC, is responsible for exploration, production, transmission, and distribution of gas. It controls six regional gas utilities. Responsibility for gas imports used to rest with Weglokoks, the former operating company of coal production and trade. Negotiations were, however, mainly carried out by the government because gas contracts were incorporated in the old trading and barter arrangements of CMEA. The division of responsibilities between POGC and the government is still not very clear. The Ministry of Industry, for example, exercises control over its management and activities. Currently, the government is moving towards a restructuring of the gas sector, aimed at introducing more commercial criteria into operations. At a later stage, the gas sector may be opened for private capital.

An important issue facing the Polish gas industry is the choice of source of new supplies. New supplies are needed for co-generation and also to provide a substitute for town-gas; both end-uses are presently dominated by coal. Russia is the sole source of imports. Other supply options, including supplies from the North Sea, are being considered; but on a full-cost basis, additional Russian gas volumes are probably the least costly and the only source realistically available to Poland within the next decade. Only in the longer term do gas supplies from the North Sea appear to be a realistic possibility. Additional supplies from Russia could be delivered to Poland through a new pipeline planned from Russia to Germany across Belarus and Poland.

Together the Czech and Slovak Republics are the largest users and importers of Russian gas in Eastern Europe. These two countries (the former Czecho-slovakia) have only minor indigenous reserves of gas. They have an important position because of their role as transmitter of all Russian gas exported to Western Europe. Payments by the Russian exporter for transit are made in the form of gas. Each of the two Republics has its own national gas import and transmission companies, which also own the local distribution companies. The company owning and operating the transit line to the West has been divided between the Republics. In both, foreign companies have been invited to invest in the gas sector.

Gas consumption is split between power generation (37 per cent), industry (25 per cent), and the residential sector (41 per cent, all 1989 figures). As in Poland, the two Republics are considering how to diversify their supply. New gas volumes are sought to substitute gas for coal in industry and in district heating. In the near to medium term, however, additional volumes are most likely to be imported from Russia, for reasons of cost.

Hungary imports a little less than half of its gas requirements from Russia. Gas represents 31 per cent of total primary energy consumption, close to the prevailing share in the major gas-consuming West European countries, and much higher than in Poland and the Czech and Slovak Republics. This relatively high share reflects the fact that already in the 1960s gas was being used from the country's own reserves. The development of gas use in Hungary was therefore in line with the general policy of the CMEA countries to achieve self-sufficiency as far as possible and as a priority to develop their own energy resources.

Compared to the other Central European countries, Hungary uses a large share of its gas in the residential sector. Including the gas consumed in district heating plants, residential/commercial use accounts for 63 per cent of total gas use. Direct use in the residential/commercial sectors amounts to 40 per cent of total gas consumption. Penetration of gas in the residential sector is high and increasing rapidly, with 35 per cent of households supplied with piped gas.

5.3. Conclusions

Apart from the UK, the structure of the European gas industry has been remarkably stable during the last twenty-five years. The rapid growth in gas use has not changed the organization of the market. On the contrary, it has probably reinforced the status quo. The industry remains dominated by a handful of large transmission companies which have control over imports and long-distance pipeline transport. It is less important whether these companies are public, semi-public, or private. The crux of the matter is that they are all party to an intimate social contract with their respective governments. They supply gas from foreign (or domestic) sources in adequate volumes at reasonable prices within a system that also offers security guarantees. In return, national authorities either intervene to secure their markets, if competition with other sources of energy looks threatening, or refrain from opening up markets to competition from other energy industries or from within the gas industry itself.

In spite of increasing pressures for change, this deal seems to be well rooted in both commercial and political interests, at least as far as the 1990s are concerned, but in the longer run this structure also faces serious challenges.

1. Technological development makes gas use on a smaller and smaller scale economically attractive, especially in power generation. As a result, an increasing number of independent gas users will demand better access to the grid, gradually increasing pressures from below on the transmission companies. New technologies also make gas consumption increasingly efficient, in terms of energy produced per volume of resource input. This is spurred both by economic incentives, and by environmental considerations. If climate change becomes a priority issue in the next decade, this development will receive further stimulus (and probably large public R&D funding), which could have major implications for gas use in the long run.

2. If gas consumption keeps increasing in line with present forecasts it will sooner or later face major supply constraints. With today's prices it will be difficult to bring to the market major new supplies in time to meet the expected demand increases. Growing prices will spur competition with coal and oil, and perhaps with biomass in some countries, for power generation and heating.

In any event, the expected increase in gas demand is bound to create major political concerns for future security of supply, given the precarious situations in Algeria and Russia, which could have devastating effects on the entire European gas industry.

6
Energy-Sector Developments: Electricity

The electricity sector remains the most tightly regulated energy sector in Europe. The existence of natural monopolies in transmission and distribution calls for regulation to prevent abuse of a dominant market position. Moreover the electricity sector has generally been an important political instrument for industrial and technological developments or strategies to diversify energy requirements and secure markets for domestic energy resources. Still, the regulatory systems across Europe reveal some significant differences, rooted in national characteristics of corporate traditions, the emphasis on local or central governance, and the national resource base. Despite the diversity in systems, the electricity sector showed a great degree of institutional and regulatory stability from the early 1950s, when most of the structure was established, until the late 1980s. Pressure for change in the regulatory systems is, however, now apparent in most countries. The nature of this pressure and the political and industry responses will differ from country to country. Undoubtedly, the past stability of the electricity sector has been broken, but it is unclear whether the direction of change will prove uniform across Europe. It will depend, among other things, on the resilience of existing national patterns. In looking at the future of the electricity sector, it is important to understand the factors that have shaped and confined past and current structures. Hence a relatively detailed account of key features of the electricity sector in important European countries follows, along with a summary of the main factors currently driving sector reforms and the forces working to preserve present structures.

6.1. Institutional and Regulatory Features

Historical Background

One reason for the high degree of regulation is the existence of natural monopolies in transmission and distribution, which have the potential for abuse of a dominant market position. Another reason is the strategic importance of electricity for economic development and the public concern for stable and reliable electricity supplies. Furthermore, the electricity sector is in itself a key energy-consuming sector. Many governments have therefore controlled the fuel choice in electricity generation as an important element of overall energy and environmental policies. In countries with significant indigenous energy resources, the electricity sector has been the prime target of policies favouring domestic energy production, such as for coal in Germany and the UK, and for gas in the Netherlands.

The policy objective for the electricity sector has been to provide adequate supplies at reasonable cost. Most countries have succeeded in this objective. In Western Europe electricity production increased fivefold from 1950 to 1975. Electricity prices fell continuously over the same period, reflecting technological progress, improvements in operational efficiency, and declining fuel prices.

The turbulence of the 1970s, however, disrupted this trend. The costs of electricity production started to increase, not only because of higher fuel prices but also due to higher construction costs for new plants. Together with lower economic growth this caused a notable deceleration in the growth of electricity demand: from 6.5 per cent per annum between 1950 and 1973 to 2.5 per cent from 1973 to 1983 (IEA, 1985). The response to higher oil prices was a massive conversion of generation capacity away from oil, and continued net additions of non-oil-fired capacity despite low demand growth.

The events of the 1970s did not call into question the regulatory system of the electricity sector. If anything, they led to a strengthening of government influence and control as energy security considerations emerged. The electricity sector became the prime tool for achieving diversification in energy supplies (see Fig. 6.1). An active policy was pursued to increase the use of coal in electricity production: from 1970 to 1992 coal-fired power plants increased production from 500 TWh/year to 800 TWh/year. As has been discussed above, the development of gas-based electricity was also subject to heavy-handed energy policy, primarily through a ban on gas use in new power plants within the European Community. Hence the share of natural gas in power production fell from 10 per cent in the early 1970s to 6 per cent at the end of the 1980s.

Nuclear power began to play a key role in substituting oil, notably in France but also in Belgium, Sweden, Germany, Spain, and the United Kingdom. Electricity production in nuclear plants increased rapidly from 1975, but growth fell sharply in the late 1980s in most countries. The public debate over health and safety risks that followed the Three Mile Island accident in 1979 and the

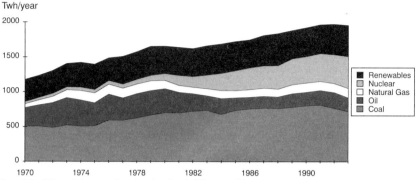

Twh/year

FIG. 6.1. Electricity production by fuel, Western Europe

Chernobyl disaster in 1986 made it increasingly difficult to find sites which met technical standards and public acceptance. This resulted in longer gestation periods for nuclear projects in most countries, with the exception of France. Moreover, some countries conducted referenda and legislative actions which put a halt to further nuclear-power developments, and in some cases even led to early closure of existing plants.

The radical shift in fuel use in West European electricity generation, with nuclear holding one-third in 1995, is a product of the institutional and regulatory forces that influenced the sector. Without governmental initiatives and commitment to the nuclear research and development programmes, the rise of nuclear power would never have happened. The tight regulation of the sector, together with market protection, have prevented many enterprises from bankruptcy in those cases where excess generating capacity and high capital costs would otherwise have led to financial ruin.

The Regulatory Systems and Ownership

Despite stringent regulation of the electricity sector throughout Europe, the actual institutional framework varies widely from country to country. In Italy, France, and the United Kingdom there existed throughout the 1970s and 1980s vertically integrated and state-owned national monopolies (UK has since privatized and broken up the vertical structure). However, direct state control and vertical integration is in fact not the typical European model. In most of Europe, electricity generation and distribution are undertaken by a large number of entities, a majority of which is owned by municipalities and regional authorities. Private ownership is also common, especially in electricity generation. Formal and informal links between enterprises, both along the vertical chain from generation to distribution and horizontally at each level, vary considerably, as does government involvement. To give a better understanding of the main features of the electricity structure we now present some country-specific characteristics.

Germany

The German electricity sector is highly decentralized, with several large generators and a large number of distribution companies. The system is closely regulated through a network of private contractual arrangements protected by Federal legislation. Nine inter-regional enterprises (*Verbundunternehmen*) dominate electricity generation, which own more than 80 per cent of the production capacity. In addition they own and operate the transmission system. The inter-regional enterprises have established demarcation contracts which map out the supply areas for each enterprise, ensuring that no competition among them arises. The system of demarcation contracts dates back to 1908 and has since been supported by the Energy Law of 1935 and the Cartel Law which granted exemptions for electricity and gas.

The inter-regional enterprises supply some of their production directly to large industrial customers, but most electricity is sold to about seventy-five regional enterprises and to some 600 local utilities. The regional companies have some production of their own and their operations are also regulated by demarcation contracts. Furthermore, they are in most cases owned or controlled by the inter-regional companies. The demarcation contracts are backed up by concession contracts between municipalities and distribution companies. Utilities are granted exclusive rights to supply customers within a local area, in return for an annual concession payment to the municipality. This payment varies according to the characteristics of the area, but is typically 5 to 7 per cent of the sales value to small consumers. For many local authorities, this represents an important income.

Legislative reform in the late 1980s changed the system of demarcation and concession contracts to some extent. Their duration has been limited to twenty years, and both types of contracts are to expire on the same date. The intention of these changes was to create more competition in the sector. Further changes, perhaps resulting in a complete removal of demarcation agreements and exclusive concession contracts, were considered by the Ministry of Economic Affairs in 1994, but did not gain full support from the Federal Government.

The owners of the eight inter-regional companies of the former West Germany are partly municipalities and Länder and partly private. Recent years have seen a concentration in ownership. In particular RWE—with a mixed public and private ownership and by far the largest electricity enterprise with one third of total German sales—has gained strength and market shares through the acquisition of regional and local companies. Moreover, RWE together with PreussenElektra and Bayernwerk, two other large inter-regional companies, jointly own 75 per cent of VEAG (Vereinigte Energiewerke AG), the production and transmission conglomerate of the former East Germany. The remaining 25 per cent is owned by four smaller West German inter-regional companies. The German Government accepted a dominant position of the three large Western enterprises in exchange for their commitment to upgrade the Eastern electricity system to modern standards, and to improve the environment by investing in abatement technology and new generating capacity.

Construction of new generating capacity is subject to a licensing procedure administrated by state governments. According to the procedures, investment projects are to meet criteria related to environmental protection and safety; the need for additional capacity must be justified, including the trade-off between capacity expansion versus energy conservation. It has become increasingly difficult to find acceptable sites not only for large new power plants, but also for transmission lines. Most new large plants are constructed on existing sites.

Electricity tariffs for small consumers are regulated by the government, whereas prices to large industrial customers in principle are negotiated between the utility and the consumer and are generally not published. The tariff regulation for small consumers is based on a cost-plus principle. Electricity prices for industrial users fall under the general monopoly legislation, which prohibits abuse of dominant market position. It has, however, proved practically impossible to contest industrial electricity prices in court because of difficulties in establishing unequivocal price comparisons between customers.

As mentioned above, West German utilities were obliged until 1995 to purchase German-produced coal at prices well above international steam-coal prices. A part of the additional cost was financed by a levy on all electricity sales (coal levy, or 'Kohlepfennig'), but this arrangement was in 1994 ruled unconstitutional by the German Constitutional Court (see Chapter 4). As of 1996, the coal levy will be replaced by a system where direct support for the coal mines is provided from the Federal budget and from the states in which hard-coal mines are located. As a consequence, electricity prices are likely to decline by about 8.5 per cent.

France

The institutional structure and regulatory system of the electricity sector in France contrasts sharply with that of Germany. France has no tradition of strong regional governance, so the state-owned and vertically integrated EdF (Electricité de France) has been able to pursue its goals without interference from local or regional authorities. Ever since it was established as a legal monopoly for electricity supplies in 1946, there has been a close relationship between the central political authorities and EdF. Its senior management has largely been recruited from the Ministry of Industry and Commerce, which is the regulatory body of EdF, and EdF employees have occasionally moved to senior positions in the Ministry. Furthermore, the top management is often replaced following a change in government. Key ministries, such as finance or industry, are all represented on the board of EdF by senior civil servants, together with representatives from other state-owned enterprises. This has the effect of securing the political loyalty of the management of EdF to the overall strategy of the government, and the boundaries between regulatory functions and commercial decisions have been less clear than in most other West European countries.

Within EdF, trade unions have developed into a significant force which cannot be ignored either by the management or the government. As in other French state enterprises which are protected from competition, such as telecoms, postal services, or gas supply, trade unions generally oppose deregulation and privatization which may eventually challenge their power and reduce the workforce. These factors partly explain EdF's—and the government's— reluctance to accept initiatives of the European Commission to liberalize the electricity market.

EdF is often presented as 'a state within the state' with its own agenda beyond political control. Its influence on energy policy issues has been considerable, but there has been practically no political controversy over EdF's role or important dispositions made by the company. Moreover, the Ministry of Industry and Commerce, which is responsible for energy issues, is directly involved in investment decisions and tariff issues. Since 1970 the Ministry and EdF have developed a 'Contrat du Plan' every five years, which establishes detailed investment and financial plans and targets, as well as guidelines for tariffs. Since the mid-1980s the Ministry has not allowed EdF to increase prices in line with inflation, despite the increased debt burden of EdF caused by France's ambitious nuclear programme. The result has been an accumulation of debt which in recent years has been reduced due to better utilization of the large capacity of nuclear power and a slow-down in the nuclear construction programme.

The vulnerability of EdF is not only confined to its financial burden, but also, as has become evident over the past few years, to the configuration of its power plants. Several nuclear plants in the 1300 MW series have developed identical technical deficiencies due to welding faults in pressurized seams and deformations of steam generator tubes. EdF's strategy of constructing nuclear power plants in series with the same technical design has clear advantages in holding down the construction costs. It is, however, also a risky strategy if design faults necessitate repairs which could affect a major part of the entire nuclear park. In addition, consecutive years of drought in the late 1980s and 1990s have obliged EdF to curtail the utilization of certain nuclear plants where river flows reduced the cooling capabilities.

In France too, there has there been opposition to the nuclear programme (see Nelkin and Pollak, 1981). Recently EdF has met strong local opposition to its plans for the siting of underground storage facilities for high-level radioactive waste. However, on a political level and nation-wide, nuclear power is accepted as a cornerstone of energy policy. More generally, environmental questions have not ranked high on the political agenda in France, where the centralist administrative and political powers have facilitated the nuclear programme. Topographic advantages also play a role: compared to other European countries, in France it is relatively easy to find river sites in thinly populated areas. As a result, however, transportation distances in France from the plants to the major consuming centres are relatively long and costly. There is growing public opposition to the construction of new transmission lines, in particular for exports of power.

Italy

On the face of it, the Italian electricity sector looks much like that of France. The state-owned Ente Nazionale per l'Energia Elettrica (ENEL), formed through nationalization in 1962, enjoys a monopoly in transmission, import, and export of electricity and is the principal electricity generator and distributor (85 per cent of total generation). The linkages between ENEL and the central political and regulatory authorities are as strong as in France. Senior managers occasionally hold positions simultaneously in ENEL and in government office. Tariffs are set centrally by an inter-ministerial committee. Tariff increases have often fallen short of ENEL's requirements to attain financial solidity, however.

Despite the apparent similarities with France, the differences are distinct.

ENEL depends on a weak and unstable central government. As a consequence, it has met severe obstacles to building new capacity over the past twenty-five years. It has been especially difficult to construct new coal-fired plants, which increasingly became a target for local resistance. In Italy, most new plants are planned along the coast, which is relatively densely populated. Another important factor behind the lack of investment in generating capacity has been ENEL's difficult financial situation and shortage of capital. The closure of the country's nuclear plants in the late 1980s further aggravated capacity shortage. The result has for years been a persistent deficit in generating capacity and reliance on substantial imports from neighbouring countries, notably Switzerland and France. For fuel use in electricity generation, the result has been a continued high reliance on oil, unlike the structure of any other European country.

Italy's good luck—or perhaps the reason for not acting on these problems—is that France has had surplus capacity. Moreover, the collapse in oil prices in 1985–6 improved ENEL's financial position, since so much of its generating capacity was based on oil or gas. ENEL's lack of capacity to supply domestic demand has also encouraged a relatively large auto-production of electricity (and heat) by industrial enterprises. Independent power generation, often owned by industries, has been favoured by attractive tariffs for power sold to the public grid. All new independent power generation in Italy uses natural gas.

A decision to privatize ENEL has been close to being taken for some time. However, the issue has been repeatedly postponed, due to political turbulence. In 1993, the government authorized price increases, aiming to improve ENEL's financial position and make the company more attractive for potential buyers. The government faces severe pressure to cut its budget deficit in order to meet the convergence criteria for Economic and Monetary Union and is likely to continue its privatization programme of the energy industry, which began in 1995 with the partial privatization of ENI (see Chapter 3), and to float parts of ENEL's share capital during 1996.

United Kingdom

Prior to the privatization of the electricity sector in 1991, generation and transmission in England and Wales were undertaken by the Central Electricity Generating Board (CEGB), while distribution was carried out by twelve Area Boards. In Scotland and Northern Ireland, production, transmission, and distribution were carried out by vertically integrated companies. CEGB's structure was established in the 1947 nationalization of the British electricity industry, together with other key sectors, such as gas, coal, and steel. As in France, the nationalization was part of government policy in the aftermath of the Second World War. Close commercial links with government monitoring were established between CEGB and British Coal for the supplies of coal on firm contracts for electricity generation.

The first attempt to stimulate private investment in electricity generation came with the 1983 Energy Act, which established the legal basis for third-party access to the CEGB grid. For the first time since 1909, other producers than the Area Boards were permitted to supply end-users directly. The reform failed, however. Production from independent producers did not increase, in part because of excess generating capacity in both England and Scotland. The market imbalance was further aggravated by the decision to link the United Kingdom with continental Europe, making it possible to import relatively cheap electricity from France. The ties between CEGB and the Area Boards were strong, and this, together with a tariff structure which was insufficient to provide private investors with adequate returns, discouraged private power producers from investing in new generating capacity. One lesson from the failed 1983 Energy Act was that vertical integration of the electricity sector would have to be broken up if competition from independent generators was to materialize.

The objective in the full privatization of the electricity sector in the United Kingdom was to achieve competition, primarily in electricity generation. However, the motivation to break up CEGB also stemmed from the Conservative Government's intention to reduce the influence of coal mines on the power industry, by increasing the scope for natural gas, which has proven to be the preferred fuel choice of the new players that have been created. CEGB was split into three generation companies, the private companies National Power and PowerGen as well as Nuclear Electric, which was established as a state-owned company comprising all nuclear power stations, since it was not possible to attract private investors to take ownership of these plants. In 1996 the UK Government proceeded with the privatization of the nuclear industry. The youngest nuclear generating plants will be privatized while the state retains the oldest plants with their liabilities. The transmission part of CEGB has been established as a separate company, the National Grid Company (NGC), owned by the twelve Area Boards which were renamed Regional Electricity Companies (RECs). They are responsible for distribution of electricity, though they do not have a monopoly in that function.

The RECs have within their geographical areas the right to supply small consumers with electricity. The threshold was originally set at 1 MW, but was lowered in 1994 to 100 kW and will disappear completely by 1998. Consumers with an electricity load higher than the threshold have the right to seek supplies from another REC or any other supplier. In order to protect the RECs from the effects of competition, National Power and PowerGen were restricted from obtaining more than 15 per cent of the sales in any REC's area until 1 April 1994. After this date the limit was relaxed to 25 per cent and subsequently it will be removed entirely.

The new structure is intended to stimulate competition in power generation between National Power, PowerGen, supplies from France and Scotland, and from independent generators. Another key element of the reform is to encourage new independent generators in order to reduce the dominant position of National Power and PowerGen. The RECs have the right to produce their own electricity in separate production companies, on the condition that these supplies do not exceed about 15 per cent of their total electricity needs. This right has been used by the RECs to construct gas-fired combined-cycle gas turbine plants. Within their franchise areas, i.e. for the part of the market where there is no competition, the RECs price their electricity on a cost-plus basis under control of the regulatory authority, OFFER. Power from their own production provides the RECs with price security compared to the fluctuating prices from the pool (see below). In addition, the RECs obtain a stable and secure rate of return on investment in own power generation. On the other hand, this may have given the RECs an incentive to construct more gas-fired capacity than is actually needed.

In the market, prices are determined in the pool. The two major generators, PowerGen and National Power, are supposed to compete in the pool in order to maintain prices close to the marginal costs of production. They have, however, been criticized for manipulating pool prices to higher levels than justified by their costs. In early 1994, OFFER agreed with PowerGen and National Power to cap the pool purchase price. It is feared that this may result in cancellations and deferrals of new independent power plants, which in turn may hinder competition.

The lessons to be drawn from the unique UK privatization experiment are far from conclusive. Protection of the business of privatized companies, for example the RECs, was intended to obtain a high price in the sale of the companies to the public. The protection has apparently also created distortions in the market, which are difficult to control through a public regulatory body with heavy-handed regulation. Moreover, during 1995 and 1996 a series of mergers took place that involved the RECs and power generators. National Power, PowerGen, and Scottish Power have all taken control of some of the RECs. This is a step towards verticalization which represents a break with a distinct feature of the initial reform which separated power generation from distribution. In 1996 the government intervened and prevented agreed mergers between the generators and some of the RECs. This political step was made *ad hoc* and against the advice of the MMC. There appears to be a need to formulate a clear policy for a closer regulatory oversight of both generation and distribution.

The Netherlands

After the United Kingdom, the Netherlands was the second European country to take steps to reform its electricity sector. A comparison between the UK and the Netherlands is interesting because of the major differences involved. The structure of the electricity sector in the Netherlands was from the outset quite different to that of the UK, and the political motives for the reform in the Netherlands were not so much to stimulate competition as to improve central government control over the sector, to rationalize and strengthen the industry, and to improve efficiency and management.

Before 1989 the Dutch electricity supply industry was highly decentralized. Sixteen generating companies delivered 90 per cent of all electricity to more than seventy distribution companies. These companies were entirely owned by local and provincial authorities and were fairly tightly controlled by their owners. The reorganization led to the creation of one transmission and coord-inating company, SEP (Samenwerkende Electriciteits-Produktiebedrijven), which is owned by four electricity generators. Distribution functions are divided between forty or so companies.

SEP, which had only a coordinating role before the 1989 reform, has had its position strengthened. The company is responsible for a power-pool system where it purchases power from the generators on a merit-order basis and sells it back to the generators. Distribution companies can, according to the law, purchase power from any of the four generators within the country, but due to the alignment of prices through the pool system they do not have an incentive to seek agreements with generators outside their own district. The coordinating role of SEP also includes purchase of fuels for the generators, with gas purchases from the Dutch Gasunie of primary importance. Moreover, SEP is responsible for developing plans for adding capacity to the system in electricity generation and transmission. These plans and other parts of SEP's operations, including negotiations for fuel purchases, are carried out in close consultation with, and are eventually endorsed by, the Ministry of Economic Affairs.

Although the entire electricity sector, including SEP, is owned by the municipalities and the regions, they have only limited influence on SEP.

The 1989 reform strengthened the distribution companies, and subsequent mergers between them have further reinforced their position. In the process, however, they also want more independence and influence on the generation side and on imports. According to the legislation, imports are reserved for SEP; but this import monopoly is in contradiction with EU rules and is under attack not only from inside the Netherlands but also from the EU Commission. Some distribution companies have by mergers obtained owner control of the electricity generation company in their area. A possible outcome could be that the distribution companies would reinstate a vertically integrated system, contrary to the intentions of the 1989 electricity law, and that the central control of the system, now exercised by SEP and the government, will fall apart (ECON, 1993). Distribution companies, which are legally prohibited from large-scale generation, have started to do so in conjunction with manufacturing industries, such as chemicals. This policy was initially supported by the government, which promoted the construction of gas-fired combined heat and power plants for environmental reasons. Since 1992, this has lead to a gradual build-up of decentralized capacity, which is neither approved nor dispatched by SEP, leading to a significant capacity surplus and further destabilization of the power system.

The conflicts in the system have led the government to propose a restructuring of the electricity sector to introduce more competition and less central planning. The main innovations are the opening of the transmission grid for TPA and freedom to import and to produce electricity. The government proposes a cautious step-by-step liberalization. It is important to avoid financial losses on the part of producers stranded with overcapacity. In fact, the government proposes that full-fledged competition will only be introduced when the current overcapacity is absorbed. The government's analysis points to an absorption of the surplus around 2003–5.

Norway

Norway is of particular interest because the country has embarked on a reorganization of its electricity sector which is far more radical than in any other European country. The electricity system, totally based on hydropower, has traditionally consisted of a large number of companies. Prior to a new Energy Law, which was made effective in January 1991, there were some 70 generating companies and 230 distribution companies. The state-owned company Statkraft has been the dominant generator, with 30 per cent of production. In addition, regional and local authorities have major owner shares in production and distribution, making the total public owner share of production and distribution some 75 per cent of the total. The remaining 25 per cent is primarily owned by manufacturing enterprises (Hope *et al.*, 1993).

A few years before the Energy Law came into effect, various proposals were mooted for a totally different kind of reorganization. One plan was to create twenty regional vertically integrated companies. This plan aimed at rationalizing the sector, as there should have been a potential for capturing economies of scale through mergers. However, plans for concentration and integration were abandoned and gave way to the radical programme for deverticalization.

Key elements of the 1991 reform are:

- The transmission part of the state-owned Statkraft was separated from power generation. The generating part was formed as a stock-holding company Statkraft (still 100 per cent state-owned), whereas transmission functions are now undertaken by a separate state company Statnett, the Norwegian power grid operator.
- Common carriage was introduced throughout the grid and for all end-users. The owner of the grid is obliged to transport electricity for any party at non-discriminatory and published tariffs. The principles for the determination of transport tariffs are set out in the law. If more capacity is needed to meet the transport requirements, the grid owner is obliged to make the necessary investment.
- The market is administrated by Statnett under the auspices of the regulatory bodies (the Norwegian Water Resource and Power Board and the Price and Competition Directorate, which are supervised by the Ministry for Energy and Industry). Currently the pool consists of three markets: a spot market, a contract market, and a 'regulation market', the latter giving Statnett the right to direct or reject supply at short notice in order to balance the system.

It is interesting to note that the reform did not contain any plans for company mergers. Any movement towards owner concentration in production and distribution will be decided by the individual companies. So far, only a small number of mergers have taken place, but some of the major companies have established joint-ventures that have received differing treatment from the competition authorities, some being accepted, others being prohibited. Unless a strict policy on mergers is followed, a gradual development towards increased owner concentration in power generation is likely, with fewer changes in distribution.

Some aspects of the new system are controversial, including the speed with which it has been implemented, but the 1991 law has survived the first turbulent and experimental years. The reform was launched at a time of considerable temporary surplus capacity in the Norwegian system, caused primarily by climatic conditions. However, the main reason why prices fell sharply was that the system of a common dispatch between producers was maintained with the neighbouring countries. Within the Nordel cooperation agreement, which includes Finland, Denmark, Norway, and Sweden, power was traded between producers at variable production costs. Thus, the price in Norway was heavily influenced by fuel costs in the still monopolized markets in neighbouring countries, and less so by the balance between supply and demand in Norway. Prices fluctuated between zero and two US-cents per kWh in the first years after the reform.

Soon after prices dropped, there were fears that many companies would run into major financial difficulties due to large capital costs, or purchase obligations at high pre-reform prices. This could, in turn, have put some municipalities in a difficult financial situation. However, economic losses primarily hit large companies with considerable financial reserves and these managed to survive until prices firmed up at the end of 1993. Thus, the reform process was not derailed.

The low prices shortly after the introduction of reforms led to the shelving of virtually all investment in new production capacity. During the summer of 1996 production capacity proved to be too small, and even though imports increased strongly, prices rose dramatically. It remains to be seen whether this experience will undermine political support for the liberalized system.

Undoubtedly, the reform has on a broad scale stimulated a mentality of economic efficiency, but some major issues are still unresolved. Not surprisingly, transport tariffs cause conflicts and are to some extent an obstacle to economic efficiency in the market. However, these problems are possibly easier to overcome in Norway than in many other countries due to the long experience of fairly complex transmission and distribution arrangements which has characterized the Norwegian hydropower system.

As of 1 January 1996, the Swedish electricity market is deregulated along much the same lines as in Norway. A common Norwegian/Swedish electricity bourse, Nord Pool, has been established, leading to a convergence of Norwegian and Swedish prices. Limited transmission capacity leads to periodic price differences between the two countries. Finland also deregulated its market for medium-sized and large customers in late 1995, and has set up a separate bourse. An integration with the Norwegian and Swedish market will probably take place at a later stage, but first a couple of obstacles have to be removed. The control of cross-boundary transmission lines is due to be transferred from the dominant power company IVO to a new independent company. Energy taxation, which today puts Finnish producers at a disadvantage, will probably have to be changed.

Also important is trade in electricity with still tightly regulated Denmark and Germany. The interface between a market-based hydro-system in Norway and monopolized thermal-based sectors in other countries raises commercial and energy policy questions.

Central Europe

Forty years of central planning and Communism in Central Europe have left behind an electricity sector widely different from that required by a market economy. Even though state control and regulation of the electricity sector are common also in market economies, they have been nowhere near as pervasive and devastating for economic efficiency as in the former Communist countries.

The institutional patterns of the electricity sector were determined by historical traditions in Central Europe, with a multitude of entities in Poland and one large administrative unit in the former Czechoslovakia. Common to all countries was government control of all investment decisions and important operational aspects. Though technical expertise is quite high throughout the region, lack of managerial and financial autonomy and accountability, as well as the lack of modern management techniques, has led to a massive waste of resources in electricity supply. Today, the electricity sector is in a period of transition, moving away from the centrally planned system, where the main responsibilities of the state enterprises were construction and engineering. The electricity sector is faced with some difficult challenges: (i) electricity prices are generally too low to cover the costs of generation, and politics obstruct real-price increases; (ii) the average age of the power plants is high, and there is a need for investment in upgrading and modernization; and (iii) new environmental regulations call for substantial investment.

In the centrally planned system all energy prices were kept low and had no relation to costs. After 1989, electricity prices were increased, but the real increases were most important in the early phase of the economic reform process in 1990. Later, unemployment, declining real incomes for parts of the population, and government anti-inflationary policies have generally prevented further increases in the real level of electricity prices. In addition, the electricity sector is suffering financially from unpaid bills of other state-owned enterprises. Inflation has eroded the book value of the assets of the electricity companies and thereby the value of depreciations for tax purposes. Although electricity prices are lower than costs, generation companies often have a taxable nominal profit. Tax payments further undermine the financial viability of the companies.

The average age of generating plants is high, and technical standards have suffered from lack of maintenance. New environmental regulations will require installation of flue gas desulphurization and low NO_x burners. Because of the pricing policy for electricity, the electricity-producing companies generate insufficient cash to finance the needed investment or to guarantee potential private investors, such as electricity companies from the West, the required return on invested capital.

The precarious financial situation of some of the Central European utilities probably also implies that the improvements in efficiency and managerial responsibility—the aim of the restructuring—are not being achieved. Some companies continue to rely on government guarantees for investments, and the ultimate influence of central government on management decisions will remain strong. Thus there is a contradiction between the overall aims of the restructuring of the industry and the pricing policy, resulting in serious financial constraints.

Poland

As the largest consumer of coal, the electricity sector has a significant role in the Polish economy. In the central planning system, the electricity sector was based on indigenous energy resources, and the aim was to limit imports of raw materials from the partners in the CMEA system. Electricity is integrated with the district heating sector. District heating and combined heat and power production is important in Poland. About half of the total residential space heating is supplied by district heating.

Organizationally, the electricity sector is now in a process of a restructuring. A major objective has been to separate the enterprises from the central government in order to create transparency and clearly defined responsibilities. The state enterprises will be transformed into state-owned, but in principle financially independent, joint-stock companies. For hard-coal plants, the individual generation companies are being organized into four holding companies controlling about 10 per cent of the generation capacity each. The lignite plants will be organized in three companies owning both the generation plants and the mines. The national grid is responsible for purchasing electricity from the generating companies and reselling it to the distribution companies.

The aim of these reforms has been to create an organizational structure which can promote efficiency though management independence of the individual generating plants and financial responsibility. Over time the generation market will gradually be opened up for competition. Distribution, however, will remain as local monopolies. Private investors may obtain ownership shares in generating plants in connection with modernization or upgrading of the plants, but there are no plans for full-fledged privatization. To date, however, the entire electricity industry is still publicly owned, though many of the individual companies have been transferred into financially independent joint-stock companies, which should increase operational efficiency.

The Czech and the Slovak Republics

The generally liberal attitude of the government towards privatization and market reform is also reflected in its policy towards the electricity industry. In the Czech Republic production and transmission has been in the hands of the national monopoly CEZ (Ceske Energeticke Zavody). Several of its production units have already been sold off and some of the new companies have brought in foreign shareholders. The company is also actively seeking finance in the international markets. A minority share in CEZ itself has been sold and more shares will be distributed in the government's final round of voucher privatization, perhaps bringing the private share up to 49 per cent. Some of the eight regional distribution companies will also be privatized, but the state will retain majority stakes. The goal of the Czech Government is to privatize the competitive parts of the electricity sector, while retaining state ownership of the rest. Competition from foreign sources has gradually been allowed, since 1994. Competition will also be encouraged by opening up the CEZ national grid to third parties. The steps taken so far indicate that the Czech industry may accomplish its modernization relatively fast and that the legacy of the Communist system, as described above, can be overcome earlier than in the other Visegrad countries.

The Slovak Government decided in early 1993 to privatize the power producers, including the nuclear facilities, whereas transmission continues to be state-owned. According to the plan, production plants are to be transferred to a state holding company, from which they are to be sold, partly to foreign investors. The still unstable political and economic situation in Slovakia is delaying implementation of this plan. Electricity prices have not been liberalized and are far below levels that cover the full generating costs and allow the industry to undertake the necessary reinvestment.

In both the Czech and Slovak Republics coal (lignite and hard coal) continues to be the dominant fuel in electricity generation, covering some 65 per cent, while nuclear power covers most of the remainder. The share of coal is expected to decline further, partly due to the unprofitability of local coal and partly because the governments want to switch to cleaner fuels for environmental reasons. Both governments intend to go ahead with the expansion of existing nuclear capacity and the refurbishment of existing nuclear plants, which do not meet Western safety standards. Western expertise and capital are essential for these plans to be implemented, which cannot be obtained without a full commitment to acceptable safety standards. However, even with such a commitment, opposition in neighbouring countries, such as Austria, to the nuclear expansion plans of the two Republics is likely to continue. Such opposition has in part contributed to the withdrawal of Western sponsors from Slovak plans to complete the half-finished Soviet-type nuclear power plant at Mohovce.

Hungary

Hungarian electricity is produced using a more diverse mix of fuels than in the other three Visegrad countries. Only 30 per cent of primary energy demand is covered by coal, 45 per cent by nuclear power, and 25 per cent by oil and gas. Plans to expand nuclear capacity have been shelved for the time being. Until 1991 the whole electricity sector was organized in the state company MVM (Hungarian Electricity Board), which had twenty-two subsidiaries. From 1992, MVM has been reorganized into various production and distribution companies, and the industry now enjoys increased financial autonomy. MVM has still significant financial difficulties, as prices are regulated by the government and kept artificially low for political reasons. Distribution companies are partly owned by regional authorities. According to plans put forward by the government in early 1994, the company's operations will be regrouped in order to privatize the distribution and production sectors. Up to 100 per cent of the shares will be sold. However, earlier offers of smaller shares in the six production companies did not attract high enough bids to satisfy the state holding company now controlling them. The present government maintains that the national grid and the Paks nuclear power station will remain fully state-owned.

Partial privatization of the main production and distribution companies began in autumn 1995. It attracted significant interest from Western power companies and other investors and several of the companies were sold. Any success will depend on the government's long-term commitment towards competition, market reform, and price liberalization.

6.2. Pressure for Change, Struggle for Status Quo

End of Stability?

The institutional structure of the power sector which for most countries was formed in the late 1940s and early 1950s, remained remarkably stable up to the late 1980s. However, since privatization was initiated in the United Kingdom and the EU Commission started its push for freer trade and third-party access (TPA) in electricity and natural gas (see section 8.1, below) reforms have been launched or are on the drawing board in several countries. The electricity industry has rapidly changed in the United Kingdom, Netherlands, Norway, and in Central Europe—where societal and economic changes made a continuation of the status quo impossible. Reforms in other countries are less radical and the speed at which they will be implemented is uncertain. Another feature of institutional and regulatory changes is the diversity in the direction of reforms.

A simplified summary of the structural features and developments of the regulatory systems in Western Europe is shown in Fig. 6.2. Black dots indicate the status prior to 1990 and the arrows show the direction of change from 1990 to 1994. The x-axis of the figure measures the level of integration and competition. The y-axis indicates the degree of central control and ownership exercised by the government. In the upper part of the diagram there is a high degree

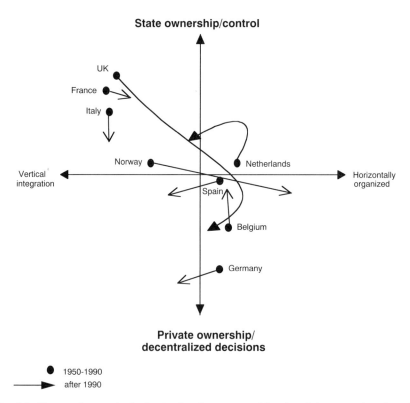

Fig. 6.2. Recent changes in the institutional structure of the electricity sector in selected European countries, 1990–1995

of central control over the entire system, whereas in the lower part companies have autonomy from state intervention to private ownership. As the figure indicates, since 1990 only the United Kingdom and Norway have definitely moved away from central control and vertical integration to a horizontally organized structure. The development of other European countries over the last four to five years (the arrows) is far more diverse. The moves towards ownership concentration and vertical integration in Germany, Spain, Belgium, and the Netherlands have been more conspicuous than the timid and not yet implemented moves towards deverticalization and commercialization in France and Italy.

In Italy privatization of ENEL is currently under consideration. If implemented it will reduce direct governmental control over the sector, but existing vertical organization is planned to remain intact. In France, the roles of the state-owned utilities in gas and electricity and the possibility of TPA is under discussion. The government and EdF argue strongly against TPA. Both the gas and electricity sectors are considered natural monopolies, where it is argued that TPA would undermine the obligation to serve and would result in inefficiencies. EdF's vertical structure is seen as a guarantee of rational planning and efficiency. The French Government would accept, however, that large consumers of electricity could be granted the right to import for their own use, but imports should be negotiated in a planned manner together with EdF. The strategy of the government is to maintain the vertically integrated companies, avoid TPA, and to open the market only to the minimum extent made necessary by EU rules and the pressure for change from the market and from other countries within the EU.

In the Netherlands the Energy Law of 1989 meant greater state control and only little competition in the market. However, counter-actions taken by distribution companies may reverse this development and thus reduce direct state control. In Germany, there has been some concentration of the industry, in particular in the new Länder. Efforts of the government to introduce competition have failed. In addition to the developments in the countries described in detail above, the institutional and regulatory trends since 1990 of two other European countries are also shown in Fig. 6.2.

In Spain a concentration with both horizontal and vertical integration around two major companies—state-owned Endesa and the privately owned Iberdrola—has been in progress for some time. The two companies have agreed on a 'peace accord', aimed at halting competition between them. The processes of ownership concentration in Spain in both the electricity and gas sectors have been actively supported by the political authorities. The background was the weak financial position of the electricity industry, as the industry had to cover the costs which resulted from the cancellation of the nuclear programme in the 1980s. Hence, Spain is moving towards vertical integration in two major companies. The future influence of central government is uncertain.

In Belgium the three major electricity generators merged into Electrabel in 1990. Through a complicated web of holding companies, the influence of a core of institutional shareholders has been strengthened and there has been a strong horizontal integration. There is also a trend towards a strengthened vertical integration. The rationale behind the reorganization was not to open the way for more government influence on the sector, but to prevent foreign companies from taking over the Belgian electricity industry. It was feared that financially strong German utilities would make inroads into the Belgian market.

More Competition—or Less?

Moves towards concentration, in countries like Germany, Spain, and the Netherlands, may have led to rationalization and productivity gains within the sector. In general, however, competition has not been fostered and it is unclear whether consumers have benefited through lower electricity prices. There has been increasing pressure for greater competition since the UK privatization was launched in 1989 and the EU Commission started its work towards an Internal Energy Market (IEM) as part of the Single Market process. Further economic integration and industrial competition, in Europe and globally, will intensify the pressure for lower electricity tariffs. Likewise throughout Europe there is a general policy drive towards deregulation which eventually may reduce the protection accorded to the electricity sector by national energy authorities.

These forces have led to the adoption of the EU directive on a liberalization of the electricity market (see Chapter 8). An important element in this directive is the negotiated TPA for transmission. In essence, TPA will allow electricity distributors and end-use consumers to negotiate supply arrangements with any producer, on the understanding that transmission capacity is available at reasonable cost. TPA would break up the monopolistic structure of electricity generation; that is why it has been met with fierce opposition by the electricity industry of most European countries. The EU liberalization directive is only a first and cautious step towards a more competitive market. Opinion on competition is not uniform throughout Europe, and there is also conflict within countries on this question. This makes it difficult to predict the future direction and speed of reforms (further discussed in Part II). The following summarizes some of the forces promoting increased competition in electricity, and the factors that work against it.

Surplus Capacity and Electricity Trade

Imbalances in capacity utilization should be a precursor to trade. However, until the second part of the 1980s, electricity trade primarily involved the transfer of electricity between utilities in neighbouring countries to level fluctuations in load and to provide mutual standby power in case of outages or faults in transmission systems. Such exchanges are often pure barter agreements without any financial transactions taking place. Planning and construction euphoria in the 1980s and lower than projected demand due to economic recession led to a growing capacity surplus, in Belgium, the Netherlands, Germany, and in particular in France.

The surplus of French nuclear capacity, together with the differentials in tariffs, put authorities and utilities in other countries under pressure to allow imports from France. This coincided with early initiatives by the EU Commission to do away with the import and export monopolies. Initially the French Government, and also EdF to some extent, were in favour of reforms which would give EdF direct access to signing contracts with foreign electricity consumers. However, the ultimate implications—a break-up of export and import monopolies in electricity—soon made the French fervent opponents of free trade in electricity.

Meanwhile, trade beyond the traditional load-balancing exchanges increased from 120 TWh in 1985 to almost 190 TWh in 1995. A major part of this increase were exports from France to Italy, Switzerland, Germany, the United Kingdom, and Netherlands. Furthermore, trade is dominated by contractual arrangements between major utilities in the countries involved, and so it does not threaten the monopolistic structure in importing countries. In the case of French exports to the UK, where EdF offers electricity to the pool, a competitive market has already been established.

Price Differences

The persistence of major differences in electricity prices between countries follows from lack of free trade and competition. The considerable disparities in electricity prices, as shown in Fig. 6.3, have emerged through a mixture of strategies pursued by power companies and by government policies imposed on the sector. Increasingly, price disparities across borders represent a pressure for sector reforms and increased trade in electricity, most visible in Germany where industry has expressed strong discontent with current price levels.

Intuitively one would expect part of the difference to be explained by France and the Netherlands having a larger share of energy-intensive industry and thus

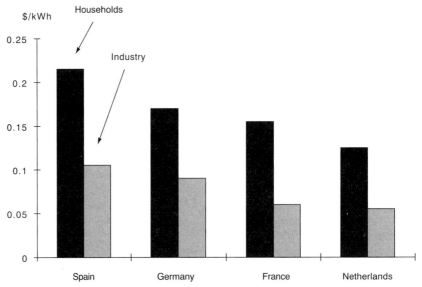

FIG. 6.3. Electricity prices in Spain, Germany, France, and the Netherlands, 1992
Source: IEA (1994*d*)

more large base-load power purchases at modest costs. However, price statistics suggest that this factor plays only a modest role in explaining the price differences to industry of electricity. According to electricity statistics published by the EU (Eurostat, 1993), large industrial consumers (70,000 MWh and 7,000 hours-load per year) in Strasburg paid only 62 per cent of comparable tariffs in Frankfurt for the year 1992. Industrial tariffs in Rotterdam were 8 per cent below the level in Strasburg. Dutch industry and households are offered electricity at about half the price paid by the Germans a few kilometres across the border, even though some 10 per cent of the electricity in Netherlands is in fact imported from Germany. Obviously the differences in tariffs for industries located very close to each other represent a major pressure for relaxation in trade restrictions.

Costs and Financial Vulnerability

Germany's high electricity prices are to a large extent caused by the high prices charged for national coal supplies, concession payments to municipalities, high costs in nuclear plants, and stringent environmental standards. Moreover, the monopolistic structure of the sector and a lenient tariff policy by the authorities have fostered a cost-plus regime with healthy financial development for the major electricity enterprises. Financial strength and technical expertise (in particular with regard to clean-coal technologies) make the large German power companies better prepared for an open and competitive energy sector in Europe than are companies in many other countries.

On the surface, the two major Spanish companies, Endesa and Iberdrola, may also look strong, in view of the operating revenues that result from their high electricity tariffs. However, their financial situation is difficult, due to the debt accumulated through Spain's nuclear programme. High electricity prices are an important element of a policy aimed at strengthening the power sector in Spain. The current financial status of the sector would make it vulnerable if consumers were to be given access to imports of low-cost French electricity.

The French nuclear programme has, as mentioned above, resulted in a substantial company debt. Unlike in Spain, EdF is capable of servicing this debt without the protection of high electricity prices. In fact, EdF has in recent years been able, through its healthy operational results, to reduce its debt. The French electricity system appears to have a cost-structure which makes it highly competitive within a deregulated electricity market in the medium term. The competitiveness of the existing nuclear plants in France does not, however, make nuclear the economic choice for new generating capacity. In a deregulated market, new nuclear capacity seems unlikely to be able to compete with coal and natural gas. This partly explains the strong French opposition to the EU Commission's proposals for electricity sector reforms.

Policy Considerations

Electricity is no longer the vehicle for industrialization and social policy that it was in the 1950s and 1960s. Favourable tariffs are still granted to energy-intensive industries in many countries, but such arrangements will inevitably shrink in importance as these industries decline and the surplus capacity in electricity generation vanishes. The electricity sector in itself is, however, still an important element of industrial policy, simply because it is one of the most important industries in Europe in terms of value added and employment. Any policy that could weaken the industry's market position or staffing—for example, through domestic or international competition—would encounter considerable resistance.

Moreover, these vested and sector interests go beyond the electricity industry in narrow terms. Again France is a prime example, with its large nuclear conglomerate vitally dependent upon deliveries to the electricity sector. Any radical changes to these structures—and not only in France—would touch upon both industrial policy in a wide sense, including research and development, and on national pride and security issues. In Germany, the change in the coal policy aligning German coal prices to international prices is a precondition for a liberalization of the electricity market. The political concerns for employment in the coal industry mean that a swift deregulation and liberalization of the electricity sector is strongly opposed. An immediate opening of the electricity market would mean that the coal industry had to reduce its capacity faster and more steeply than was intended by the government.

In the Netherlands, the government is advocating a gradual liberalization of the electricity sector. It is feared that a fast move to competition would endanger the financial position of the existing production companies. The production companies are in the current protected environment in a good financial position, but overcapacity would result in prices declining to lower than historical costs.

Technical Characteristics

The technical characteristics of electricity supply and demand are cited as arguments against competition in electricity supplies. The technical features of electricity require a close coordination of all stages of supply in order to match fluctuations in supply and demand. It is argued that competition puts security and reliability at risk. After some years of experience with the pool system in the United Kingdom and the more radical deregulation in Norway, there is, however, no evidence of serious technical disruptions to the system.

Another argument is that the regulatory requirements of the system under a TPA regime and the costs through greater commercial risks in the power sector will more than counterbalance the potential economic gains to be made through competition (Eurelectric, 1991). It is furthermore argued that the lower prices which large and powerful consumers can negotiate within a TPA regime will add to the price increases that smaller consumers will have to bear, as the entire system will accrue larger costs. These issues are still hotly debated. Most probably, it will not be possible to draw any clear conclusions before a certain amount of 'real market' experience has been gained. Hence, for the further development of energy-sector reforms in Europe, the experiments currently under way in Norway and the United Kingdom will be of particular importance.

Environmental Considerations

Environmental considerations may prove to be important factors in shaping the future electricity sector. Traditionally the regulatory systems of Western Europe have offered an opportunity to exercise control over fuel and technology choices. Less state control over operations and investments in the sector will deprive national governments of an effective tool for environmental ends. Alternatively, policies need to be agreed and harmonized on the EU level. The case of nuclear power is relevant. Nuclear is one option in a strategy to reduce air pollution and CO_2 emissions. European experience over the past few years has shown that promoting nuclear power requires a strong public presence in the electricity sector by state ownership or by regulation and protection. Private interests are not willing to take on the risk related to the ownership of nuclear plants in a competitive and non-protected situation.

Small-scale co-generation has been growing fast in some countries over the past four to five years. Such plants have environmental advantages due to their high thermal efficiency and because they often are fuelled by natural gas. The economic attractiveness of local co-generation schemes hinges on the prices offered for surplus power delivered to the grid. The interface between grid owners and co-generators is therefore important. With centralized electricity sectors, co-generation is generally insignificant, whereas countries with strong local governments and municipal utilities often have more production of combined heat and power (Lucas, 1985). Recent policies in support of co-generation in countries like the Netherlands or Denmark show the need for active regulatory steps to promote these schemes.

Concluding Remarks

The electricity sector shows a high degree of stability. Apart from the reforms at the fringes of Europe (in the UK and Norway) the basic structure is the same as in the 1950s, marked by public ownership and/or tight political control. Some governments have used this to promote their industrial policy objectives through the electricity industry. Whether the goal is to create a new industry (as in France) or protect an old (as in Germany), control over electricity supplies and markets is essential. This stability demonstrates the strength of the underlying social contract between government and industry in this field, which is based on a strong sense of mutual dependence. Governments rely on the utilities to provide one of the necessities of modern life without interruptions at affordable prices. This is crucial in keeping voters satisfied and in promoting industrial development and employment. In return, public authorities accept the argument from the industry that it needs protection from the vagaries of the market. Reliable and adequate (i.e. growing) electricity supplies require large, long-term investment which can, purportedly, only be undertaken with public support, either in the form of funding (e.g. EdF) or some kind of market security (assurance against competition from abroad or from other energy carriers).

As demonstrated above, the visible form of the social contract varies from one country to another, in particular with regard to public ownership. This reflects differences in political tradition, but does not affect the essence of the social contract and the perception of mutual dependence between government and industry. The electricity industry highlights the importance of the interaction between different parts of the energy sector in our framework of analysis. The current structure was shaped first and foremost by political priorities set by governments. Market analysis cannot explain the growth of the nuclear industry in France, nor the survival of the coal industry in Germany. The industrial complexes created by political decisions, once they are firmly established, become themselves key players in the bargaining over future policies that will determine their position in the years to come. Politics has shaped the industrial structure that conditions, if not determines, policies that reinforce the structure. This feedback explains the durability of the social contract and the structural pattern over the decades and its resilience to changing political winds. Whether it is sufficient to withstand the challenges of the twenty-first century, which is the central theme of the second part of this book, is open to question. The following issues will be decisive in this regard:

1. It has become evident that there are large hidden costs in the transmission and distribution part of the electricity industry, which are areas of natural monopoly. This is where utilities can maintain extra employment and where expenditures are to a large extent not transparent. It is possible that consumers will put pressure on these margins, and that governments may see an interest in promoting leaner and more cost-efficient transmission companies. If these forces combine to acquire political strength it could signal the beginning of the end for the vertically integrated structure that prevails in this industry.

2. Increasing electricity use has undesirable consequences for the environment, not only by raising emissions of pollutants, but also by requiring difficult choices on the supply side. It will be hard to find appropriate sites for additional power plants, almost regardless of the fuel input, in the most populated areas in Europe. As long as electricity is primarily a domestic business, this will add to the pressures on the current industrial structure and the underlying social contract, based as it is on utilities' ability to satisfy an insatiable demand growth.

3. Market pressure for non-discriminatory electricity tariffs across and within countries will increase as international competition and trade develop. Progress towards political and economic integration in Western Europe will give further impetus to these forces.

7
Environment: Achievements and Challenges

After twenty-five years on the European agenda environmental policy-making seems to have reached a point of saturation, or even exhaustion, in two respects. First, the policies formulated and the measures implemented against ecological problems from the late 1960s and early 1970s have suffered from their own success. While important issues have been solved, new and more intractable ones have arisen. The old medicine does not work on the ills of the 1990s, as they are more deeply ingrained, not only in our products, but in our lifestyles.

Second, the localized problems that spurred environmental concern a generation ago was by and large amenable to political problem-solving within national borders. The first phase of environmental policy, conducted by national agencies within these limits, has reached as far as it can go. Increasingly, governments across Europe now face the problem of 'policy leakage', whereby attempts to deal with an environmental problem at home are undermined by industries, capital, or consumers moving across the penetrated borders among nation-states.

In this chapter, we shall describe the results achieved by this first generation of environmental policy-making, and the problems that remain on the agenda for the next century.

7.1. Western Europe

Local Issues

The most important local environmental problem related to energy use is the quality of urban air, which has been an important issue since the 1970s. Health effects and reduced visibility are the greatest concerns. Although it is hard to discern clear trends, the general picture shows a slight improvement in urban air pollution, mostly early in the period. Urban air problems are caused mainly by emissions of sulphur dioxide (SO_2), nitrogen oxides (NO_x), volatile organic compounds (VOC), lead, particulates, and carbon monoxide (CO).

High concentration of SO_2 in urban air affects the human respiratory system and can harm plants and animals. The emission trends in the major countries are similar, but the turning-point varies (see Fig. 7.1). Whereas emissions of SO_2 declined during the 1970s in Western Germany and the UK, they continued to increase in Italy, Spain, and France. All countries experienced a sharp decline during the 1980s, gradually levelling off into the 1990s. The most dramatic decline occurred in France and Western Germany, but French emissions have increased somewhat since their low point in 1988, whereas Western German emissions have continued to decline. In Eastern Germany, emissions increased by more than 20 per cent during the 1980s, but have fallen considerably since reunification.

Electric power stations, residential heating, and industrial energy use account for 80 per cent of anthropogenic SO_2 emissions in Western Europe. Natural sources contribute only 10 per cent of the total, in contrast to half of global emissions. A major contributor to the reduction of SO_2 emissions has been the installation of emission control equipment and product purification. This includes desulphurization of coal and oil in the precombustion phase, combustion control measures, and desulphurization of flue gases. The use of coal with a lower sulphur content has also become more widespread. Considerable expansion of nuclear power in countries like Western Germany, France, and Sweden has been another important factor. A smaller contribution has been made by conversion from coal to gas.

Fig. 7.2 shows air-quality measurements in four Western European cities.

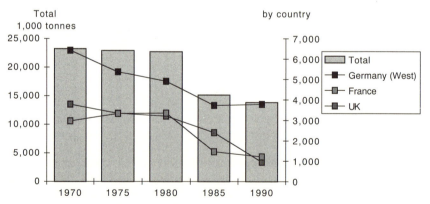

FIG. 7.1. Emissions of SO_2, Western Europe
Source: OECD (1993*a*)

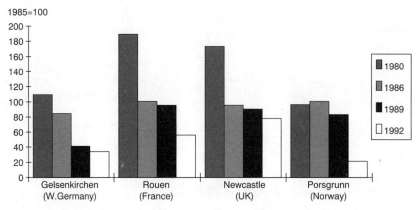

FIG. 7.2. Concentration of SO_2, selected cities (1985=100)
Source: OECD (1995)

The cities are selected on the basis of availability of adequate data and because they all had a significant number of inhabitants exposed to high levels of pollutants in 1980. (The same criteria have been used for Figs 7.6 and 7.4 for particulates and NO_x below.) The SO_2 emissions for major cities and the national level show the same trend as the selected cities below, but emissions reductions for the latter seem slightly higher. This is consistent with implementation of stronger measures in the most heavily polluted areas.

In general, concentrations of SO_2 in urban areas have decreased more than emissions. This is explained by specific restrictions on the use of high sulphur products in or near cities and the building of higher stacks at new power stations, resulting in wider dispersement of emissions and lower concentrations in urban air.

Emissions of NO_x create much of the same problems as SO_2, both regionally and locally, especially adverse health effects and damage to vegetation. National trends in NO_x emissions are quite similar in most countries. They remained stable during the 1980s after increasing significantly in the previous decade. Western Germany was able to cut emissions in later years, whereas Italy and the UK have experienced considerable growth, leaving emissions some 20 per cent higher than in 1980. As for concentration of NO_x in Western European cities, there have been significant shifts in either direction at different locations. There is no consistent downward trend as for SO_2 concentrations. Fig. 7.4 shows the development in three cities with high concentrations in 1980. Road transport is responsible for about half of the NO_x emissions, most of the remainder coming from power plants and industrial combustion. Technical solutions to NO_x emissions have been known for a long time, but

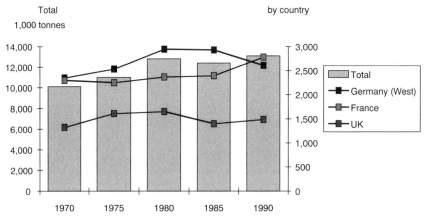

FIG. 7.3. Emissions of NO_x, Western Europe
Source: OECD (1993*a*)

only recently implemented. Policy measures have included stricter emission and technology standards.

There are several reasons why the impact of such measures has not been the same as for SO_2. Fuel substitution cannot be so easily made, for instance in the transport sector, or it has a smaller reduction potential. End-of-pipe solutions may be more costly, or technically more difficult to achieve. Furthermore,

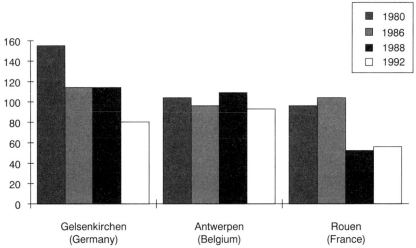

FIG. 7.4. Concentration of NO_x, selected cities (1985=100)
Source: OECD (1993*a*), OECD (1995*c*)

concern over NO_x emissions is more recent than for SO_2, leading to a delay in creating international agreements and implementing abatement measures. For example, an efficient technical solution like catalyst convertors on vehicles has only recently become widespread.

Volatile organic compounds (VOC) have direct health effects and may inhibit plant growth. Together with NO_x, they contribute to the formation of tropospheric ozone and other photochemical pollution (smog), leading to reduced visibility, adverse health effects, and damage to vegetation. Tropospheric ozone can also be transported over long distances and create environmental damage far from the source of emissions. There are no clear trends in ozone concentration, as this depends largely on weather conditions. Photochemical pollution is created in the presence of sunlight during stagnant high-pressure weather conditions. VOC comprise a wide variety of hydrocarbon and other substances, with a varied range of sources. In the OECD countries, half of the anthropogenic emissions of non-methane VOC comes from transport, and a third from the use of solvents. Emissions of VOC have been quite stable in the aggregate, increasing somewhat in Italy, Spain, and the UK.

Lead has adverse health effects. Emissions, originating primarily from vehicles using leaded petrol, were reduced by 45–60 per cent in most countries by the late 1980s, and later by even more (OECD, 1991). This is to a large extent a result of measures to reduce or eliminate lead in petrol. In several countries this has been achieved through a differentiation of excise taxes in favour of unleaded petrol.

Particulates reduce visibility and as a carrier of toxic substances affect human health. Combustion of fuels is responsible for 95 per cent of anthropogenic emissions of particulates in the European Union. The reduction in emissions, shown in Fig. 7.5, can be attributed to reduced coal burning and the installation of dust-removing equipment. After an initial fall in the early 1980s, the concentration of particulates in urban air in heavily exposed Western European cities shows no clear trend.

CO (carbon monoxide) has adverse health effects, as it interferes with the absorption of oxygen by red blood cells. Emissions are the result of incomplete combustion of fossil fuel, mainly in road traffic and industry. For Western Europe as a whole they have been quite stable in the period, although trends differ considerably between countries, as shown in Fig. 7.7. Reductions due to a shift away from solid fuels, and due to regulations on emissions from automobiles, have been partly offset (in some countries more than offset) by growth in mileage travelled.

The national differences in emission trends can to some extent be explained by variations in underlying economic growth. However, the main reason is

1,000/tonnes

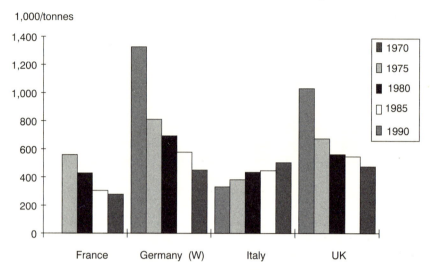

FIG. 7.5. Emissions of particles
Source: OECD (1993*a*)

differences in the scope and strength of environmental regulations, which will be elaborated below. The considerable reductions achieved in Western Germany can largely be attributed to heavy investment in environmental technology. Up to 1988, the power sector invested DEM 21 bn in desulphurization and NO_x-reducing equipment and processes, and the industry's total investment to

1985=100

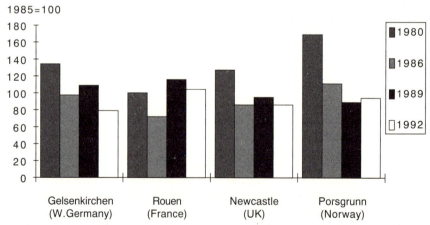

FIG. 7.6. Concentration of particulates, selected urban areas (1985=100)
Source: OECD (1995)

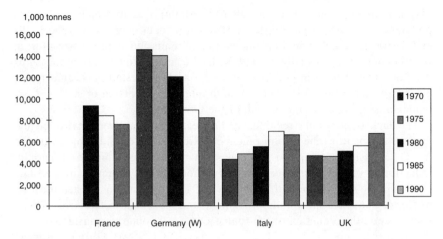

Fig. 7.7. Emissions of CO
Source: OECD (1993*a*)

reduce air pollution was DEM 38 bn (OECD, 1993*b*). This has enabled Western Germany to reduce NOx emissions from stationary sources by half since 1970, and SO2 emissions by an impressive 83 per cent between 1980 and 1990.

Cross-Boundary Issues—Acid Rain

The major energy-related regional pollution problem is acid rain, to which emissions of SO_2 and NO_x are important contributors. There are three major effects of acid rain:

- Damage to forests and sometimes to agricultural crops
- Deterioration of man-made materials, such as buildings, metal structures, and fabrics
- Acidification of lakes, streams, and ground water, resulting in damage to fish and other aquatic life

The acid rain problem came to the surface in the late 1960s, when reports of damage to forests and lakes were published. Originally seen as a local problem, the regional dimension of acid rain grew in influence during the 1970s, and regional agreements to constrain emissions were established during the 1980s. The exact nature of the link between observed damage and emissions of SO_2 and NO_x is still under discussion. Most European scientists believe the link to be strong, but research in the USA suggests that natural causes may play a dominant role. There are also uncertainties as to which of the two pollutants have the greater impact. The scientific controversy led in the 1970s and 1980s to a delay in implementation of emission control measures in several countries, especially in the UK. In Western Germany strong measures were enacted when local problems in the form of deforestation became apparent. Increasing political pressure came through the formation of strong environmentalist groups and the growth of the Green Party.

The lack of a simple relationship between emissions and environmental damage can be illustrated from some recent trends. The considerable reduction in SO_2 emissions during the 1980s has not been accompanied by a parallel reduction in observed damage. Germany's forests are apparently in the same state as in the mid-1980s, perhaps improving slightly. Data for the acidification of Norwegian lakes show no improvement, and the area affected is larger than ever, despite the reduction of SO_2 emissions in all countries considered to be the sources of most of the acid rain in Norway (SFT, 1994). For the European continent as a whole, it is estimated that more than 60 per cent of the area is subject to acid deposition that exceeds the critical loads of local ecosystems (van Aalst, 1993).

Climate Change

OECD Europe contributes 17–18 per cent of global CO_2 emissions from energy use. Emissions have been quite stable in Western Europe over the past decade. The post-war upward trend was reversed in the mid-1980s due to the economic recession and oil-price increases. National emissions do not differ greatly from the overall trend, but the Southern European countries have increased emissions somewhat more. Conversely, emissions have decreased in countries that have expanded nuclear capacity, especially France, with a fall of more than 20 per cent since the peak around 1980. In power generation emissions have been nearly halved. In Western Europe as a whole, nuclear energy led to a slight decrease in emissions from the electricity sector. Emissions from energy use in industry have fallen by more than 30 per cent since 1971, mainly due to increased energy efficiency, combined with a stabilization of industrial production. In the same period, emissions from mobile sources increased by more than 60 per cent, overtaking industry as the second largest source.

The growth in emissions has been small compared to the overall increase in energy consumption and GNP, 25 and 70 per cent respectively (1970–90). Since there are no clean-up options currently in use, the main explanation for the improved relationship between energy use and emissions is substitution

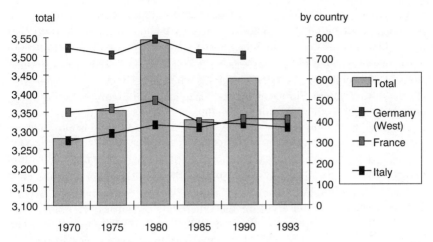

FIG. 7.8. Emission of CO_2 from energy use, mill. tones
Source: OECD (1993*a*), OECD (1995)

towards energy sources with lower or no CO_2 content, and improved energy efficiency.

4.2. Central Europe

Local Issues

Available statistics are not as good for Central as for Western Europe, but there is no doubt that the air quality in many cities and areas is much worse than in the West. Pollution is more concentrated around industrial sites and power stations, resulting in severe local problems. Recent research has established a strong link between the high level of air pollution in certain regions and the health of the population. There are strong indications that the concentration of particulates and SO_2 has fallen in recent years. This is partly due to the closure of polluting factories near cities and a ban on high-sulphur coal. Institutional and economic reforms have also increased compliance with environmental regulations, in contrast to the Communist period when measures to reduce emissions were rarely implemented. Although extensive regulations were formally in place, environmental considerations were neglected in practice. Unrealistic environmental targets still abound, however. Ambitious air-quality standards have been set, especially in Poland, but will probably not be fully enforceable for a long time, as the resources for adequate measurement and monitoring are not available.

As shown in Fig. 7.9, emissions of SO_2 per unit of GDP were fifteen to thirty times higher in Central Europe than in Western Europe in 1990, and per capita emissions three to five times higher. These ratios may be somewhat overstated, due to underestimation of GDP figures. The ratio for emissions of NO_x compared to GDP was 4:13, whereas emissions per capita were close to the Western European level. Dependence on highly polluting fuels, especially coal with high sulphur content, partly explains the high ratios, together with the inefficient use of energy and the large heavy-industry sector.

The Visegrad countries emitted some 9 million tonnes of SO_2 and 3 million tonnes of NO_x annually during the 1980s, falling considerably after the peak around 1987. Emissions from Czechoslovakia are shown in Fig. 7.10. A similar trend can be observed in the other three countries. In Poland SO_2 emissions fell by 34 per cent between 1988 and 1992, NO_x by 24 per cent, and particulates by 40 per cent. These decreases in emissions are partly due to a sharp decline in industrial production, and substitution away from coal towards nuclear power and natural gas. Some important environmental policy measures, such as a ban on highly polluting types of coal, have been implemented. Increases in energy prices have constrained demand. On the other

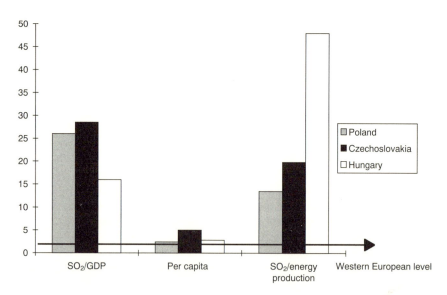

FIG. 7.9. Central European emissions of SO_2 per unit of GDP, per capita, and per unit of energy production (1990, Western Europe = 1)

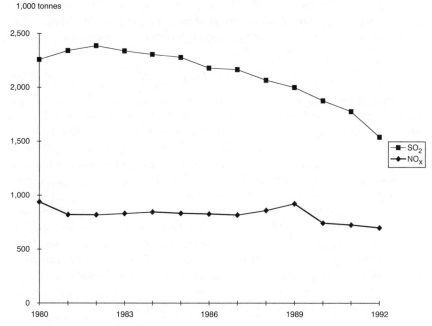

1,000 tonnes

FIG. 7.10. Emissions of NO_x and SO_2, Czechoslovakia
Source: OECD (1994)

hand, growing car use has increased emissions of CO and NO_x from the transport sector. Other emissions from the transport sector, like lead and SO_2, have been held in check by stricter fuel regulations.

Acid Rain

The Visegrad countries are both significant contributors to the European acid rain problem and suffer heavily from it. Less than 40 per cent of the national emissions from these countries end up within their own borders, but their export is to a large degree offset by emissions arriving from their neighbours. During much of the 1980s, Poland was a net importer of air pollution, mainly from Eastern Germany and Czechoslovakia. This import was not fully offset by exports to the former USSR. Poland had the highest level of deposition of oxidized sulphur in the whole of Europe after the USSR and Eastern Germany. Hungary and Czechoslovakia were net exporters by a large margin (Sandnes and Styve, 1992). The aggregate SO_2 emissions in the four countries are about half the level in the whole of OECD Europe, whereas NO$_x$ emissions are less than a quarter. The damage done by acid rain, particularly to forests, is in certain areas even worse than the damage in Western Europe.

Climate Change

High energy intensity and coal use have led to very high CO_2 emissions, compared to Western Europe. Emissions in Poland were nearly twelve times higher per unit of GDP in 1990, whereas Czechoslovakia was ten and Hungary four times higher per unit of GDP. This means that emissions per capita were slightly higher for the region as a whole than for Western Europe, even though GDP was only a fraction of the Western level. The four countries contribute some 3 per cent of the global emissions of CO_2.

Strong growth in emissions in Central Europe early in the period has been followed by a considerable decline in recent years, so that emissions are now close to 1970 levels for the region as a whole. Czech emissions fell nearly 30 per cent from 1987 to 1992, Polish emissions fell by the same rate from 1987 to 1991, and Hungarian emissions a little less than 20 per cent. This decrease is mainly due to the contraction of the industrial sector following the introduction of market reforms, and reduced dependence on coal.

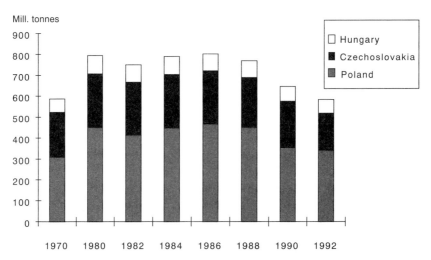

Fig. 7.11. CO_2 emissions in Central Europe
Source: IEA (1994*a*), OECD (1993*a*)

7.3. Environmental Policy

Emission reductions can be achieved through emission controls or by fuel substitution and energy conservation. Emission controls are targeted towards local and regional environmental issues. During the 1970s and 1980s emission controls have been the principal measures of environmental policy. Although the policy has affected fuel mix and energy efficiency, this has, as described in Chapters 2–6, primarily been motivated by other objectives.

This chapter reviews policy measures that directly address energy-related environmental concerns. The relevant policy instruments are: regulation and standards (command and control); information and voluntary agreements; energy taxes and emission charges; and emission trading and joint implementation.

Regulations and Standards

Regulations and standards mandating emissions control has remained the backbone of environmental policy throughout Europe. Since the early 1970s all Western European countries have used legislation and administrative measures to regulate polluting activities from energy production and use. All countries provide, for instance, a legislative framework for an air-pollution control programme. Direct regulation includes various instruments. It is useful to distinguish between quality standards, emission permits, and product standards.

Quality standards specify the character of the recipient medium, for instance air or water quality. Such standards define permissible concentrations of specific pollutants, measured in terms of contents per unit of air or water. Ambient air-quality standards are used in many European countries. Among EU member states, especially between Germany and the United Kingdom, there has been a debate about whether to use quality standards or emission standards. Discussion concerning the directive on drinking-water quality was typical of this controversy. While the United Kingdom preferred emission standards, Germany favoured quality standards. A conflict of economic interests in the two countries lay behind the conflict. The EU now has a number of directives concerning air pollution, which include quality standards, but also guidelines on emission standards for a range of pollutants.

Emission permits set the maximum quantity for allowable discharges, and may include a ban on emissions or products when a certain threshold, such as an air-quality standard, is exceeded. Permits are given by licences, and are normally combined with the use of process standards, for example the norm of 'best-available-technology' (BAT), and/or quality standards. As with quality controls, emission permits are primarily used to control concentration of pollutants in areas that are severely affected, for example, sulphur concentration in urban or industrial centres. Process standards specify the production process or the techniques to be used. The German Government's regulation of power plants illustrates the use of emission and process standards. On the basis of best-available technology, strict limits have been set; and in order to meet the requirements, power plants in Germany have to use flue gas desulphurization for SO_2 control, and catalytic reduction for control of NO_x emissions. Old power plants unable to fulfil these emission standards have to close down (OECD, 1993b). The German Government also uses such standards to regulate emissions from refineries, chemicals factories, iron and steel production and processing plants, glass factories, food production plants, and cement works. In the Netherlands, measures include regulations on large combustion installations, small fossil-fuel-fired installations, and waste incineration plants, and in controlling the sulphur content of light fuel oil.

Product standards specify quality norms for fuels, primarily petroleum and coal products. The most common product standards are for permissible sulphur content of fuel oils. In addition regulation of lead in petrol has been an important and forceful instrument in combating urban air pollution. Combined with taxation policies and 'process standards' for vehicles this has resulted in a major reduction in lead emissions.

Information and Voluntary Agreements

Information, education, and training have been applied as measures to change the polluter's perceptions and priorities. The development of voluntary agreements is an important part of this approach. Many governments have encouraged environmental efforts by companies and institutions through such means. Eco-audits and eco-labelling schemes have been widely discussed and partly implemented throughout Europe. So far, however, such measures have had little impact on local air-pollution or energy-efficiency trends.

The promotion of self-regulation is a key objective for environmental policy in the Netherlands, where environmental technology policy is based on cooperation between government and industry. In Germany, some forty voluntary agreements between industry and public administration had been established in various branches of business by 1992. In Italy, there is an agreement between the Ministry of Environment and the automobile industry regarding development of low-emission vehicles. In the United Kingdom, the Prince of Wales initiated a scheme, 'The Making of Corporate Commitment', involving 1,600 firms.

Voluntary or negotiated agreements have recently gained considerable attention as a viable climate policy instrument, particularly in industry where other instruments aimed at reducing energy use and emissions of CO_2 have proved difficult to implement (see below). Again the Netherlands is at the forefront, with agreements between industry and the authorities to reduce the energy intensity by 20 per cent from 1989 to 2000. The county is *en route* to reach this target. Germany and Denmark are also setting up formal or informal agreements with industry to improve energy efficiency as part of climate policy. Some argue that much of the agreed reductions would happen even without the agreements. Others contend that the agreements help to raise awareness of cost-free or cheap efficiency options (see the discussion of the efficiency gap in Chapter 2). With the current rate of reduction in energy intensity throughout Europe, a reduction of 20 per cent, which is envisaged for the Netherlands, appears to in excess of what could have been expected without the agreements.

Energy Taxes and Emission Charges

It is probably safe to say that emission charges played a negligible role until 1990–1 for energy prices and energy-market developments, and in the general debate about environmental policy. Although energy taxes increased markedly during the 1980s, this was not driven by environmental considerations, but rather by requirements for public funds and possibly for reasons of energy security.

However, environmental concerns have not been totally absent from taxation policies. Most Western European countries have made a tax differentiation between unleaded and leaded petrol in order to give consumers a price incentive to buy less polluting fuel, often in combination with technical standards and economic incentives to favour cars equipped with three-way catalytic converters and diesel engines with emissions under a certain level. Some countries also introduced sulphur taxes in the 1980s, normally with tax reductions being applied when control technologies are installed.

Despite these examples of environmentally motivated taxes and an increasing interest in economic instruments, practical implementation remains fairly limited. The rhetoric about market-based measures has not produced major changes in environmental policy so far (Aarhus and Eikeland, 1993). The reasons for this reluctance are largely political: if environmental levies are to have an effect on production or consumption they must be set at a high, and hence uncomfortable, level, and this can easily produce a political backlash from affected industries. Emission charges seem to suffer from lack of political support, probably because they are often seen as one more pretext for raising taxes.

One exception is the introduction of carbon taxes in the Nordic countries after 1990. It is useful to review the most important features of the tax system that have been implemented in Finland, Sweden, Norway, and Denmark. Three different aspects of the Nordic energy and carbon taxes are worth highlighting:

- There are considerable differences between the nominal and effective tax rates.
- Energy taxes were lowered when the carbon tax was introduced.
- The carbon tax is not applied uniformly across fuels.

The nominal rate is the general rate decided and denoted by political authorities as the carbon tax rate. The effective tax rate is the average carbon tax rate after it has been adjusted for exemptions and tax relief. As shown in Fig. 7.12 the effective rate is, with the exception of Finland, much lower than the nominal rate. Most tax exemptions are granted on the grounds of not losing out in international competitiveness. The nominal rates are highest in Norway and Sweden, being twice as high as the energy/carbon tax proposal put forward

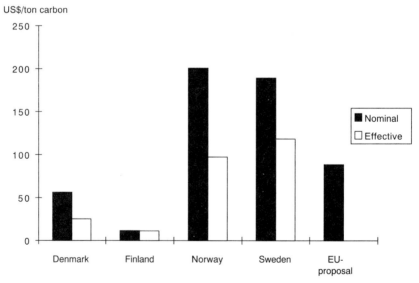

FIG. 7.12. Nominal and effective tax rates in four Nordic countries, 1993
Source: Haugland (1993)

by the EC Commission (for the year 2000). According to this proposal the tax would initially be introduced at $3 per barrel of oil equivalent (boe) in January 1993, and rising by $1 a boe every year to reach $10 boe in the year 2000, corresponding to 90 US$/ton carbon.

The effective tax rate as shown in Fig. 7.12 does not, however, give a complete picture of the additional taxation contributed by the carbon tax. The reason is that other energy taxes are often modified (generally reduced) when carbon taxes are introduced. Both Norway and Sweden have reduced and partly abolished the general energy tax on fuel oils in parallel with introduction of the carbon tax. For transport fuels, on the other hand, total taxes have increased as much or more than the carbon tax rate. The net increase in taxes on energy products has therefore been lower than the effective carbon tax level.

In principle, carbon taxes should be applied uniformly across fuels. Except for Finland, this is not the case, due to differences in both nominal rates and the various exemptions. Fig. 7.13 compares the effective carbon tax rates for coal, oil, and gas. Only Finland has the same effective carbon tax rate for all fuels—i.e. the tax rates for all fuels are fixed in accordance with the carbon content of the fuels, and no tax exemptions or refunds are permitted. The tax

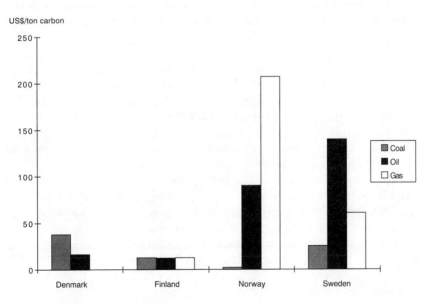

FIG. 7.13. Effective carbon tax by fuel in 1993

exemptions result in different rates by fuel for the other countries. In Norway and Sweden the effective rates for coal are very low because coal is extensively used in industrial sectors that are exempted from taxation.

Emission Trading and Joint Implementation

Tradable permit schemes can take various forms. Normally, target levels for the pollutant in question are set by environmental authorities for each point source and a system is imposed whereby polluters can trade permits to emit more or less than their individual targets. An effective scheme of tradable permits implies that emissions are abated at the lowest possible cost, thus the overall cost of reaching a specific emission target may be considerably lower than in a system of direct regulation (e.g. emission standards). Industry often prefers tradable emission permits to taxes since they provide greater flexibility, lower costs of implementation, and do not involve transfer payments from industry to government. Still, tradable permits have played a minor role in European environmental policy, unlike in the United States where such schemes have been used for SO_2.

Since 1989 similar schemes, under the name of Joint Implementation (JI), have been discussed to control emissions of greenhouse gases. JI are bilateral deals where countries with high abatement costs invest in countries with low abatement costs and receive credit for some portion of the reduction in emissions. In 1992 the United Nations Framework Convention on Climate Change (FCCC) listed JI as one of several actions to achieve the objective: 'to protect the climate for present and future generations'. FCCC states that by the implementation 'of policies and measures jointly with other parties may assist . . . in the achievement of the objective of the Convention'.

In Berlin, in the spring of 1995, the first follow-up conference to the Framework Convention declared that JI had a potential and that a pilot phase was appropriate. The pilot phase has a deadline of 31 December 1998 to determine the criteria for JI. Many European countries view JI as a promising mechanism to lower costs of abatement, and accordingly have established pilot projects to help define JI criteria, measures, and monitoring. Nevertheless, substantial methodological as well as practical issues, such as crediting, will need to be resolved before any widespread AIJ implementation.

Several European countries such as the Netherlands, Denmark, Sweden, and Norway have JI pilots at various stages of implementation. Potential JI projects cross a wide range of sectors and facilities, with projects ranging from reforestation in the tropics to district heating facilities in the Czech Republic. Relevant JI projects depend on the industrial structure of the

particular country, the stage of market development, and the regulatory structure. The latter issue is especially important if JI is to be funded by private
capital flows.

7.4. Conclusions

This chapter has pointed to the close connection between energy use and the
environment, and shows that some environmental problems are more amenable to political regulation than others. In Western Europe only one of the
urban air problems can be considered well on the way to solution, namely
lead emissions. Considerable progress has also been made in SO_2 concentrations. Of the remaining problems, NO_x concentrations seem both to create
the gravest problems and to be most difficult to reduce. For NO_x and most
other pollutants there are two counteracting trends. One is the development
and implementation of new technology and other measures to combat emission. The other is growth in the underlying factors influencing emissions,
particularly in transport.

In Central Europe some improvement has been achieved in recent years due
to fuel substitution and the decline of the industrial sector, but the situation is
still much worse than in Western Europe. There is a sizeable potential for
improvement through better energy efficiency and use of end-of-pipe clean-up
technology. Considerable improvements can therefore be made at a lower cost
in Central Europe, as witnessed in the eastern part of Germany.

Only modest measures have been implemented to reduce CO_2 emissions so
far. Carbon taxes are in place in the Nordic countries, but the proposal for
such taxes in the European Union has remained stalled. Costs of reducing
CO_2 emissions are generally high, compared to the costs involved in solving
other environmental problems. The distribution of costs and benefits of
taking steps to reduce emission of greenhouse gases is, furthermore, uncertain. Together with a lack of multilateral institutions able to cope with such a
question, this has made internationally coordinated measures very difficult to
achieve.

Experience since the 1970s thus demonstrates that (i) issues with a local
impact are much easier to solve than those with more diffuse, widespread
effects and a longer time-horizon, and (ii) effective solutions require not only
recognition and understanding of a given problem, but, more importantly, that
appropriate technical and political countermeasures are at hand. Achieving
emission reductions has proved easiest where technical solutions and add-on
pollution control are easily available. Where such solutions do not exist, or are

very costly, significant emission reductions can only be achieved through changes in energy use. This has proved a much harder task, due to the key role of energy supply and the dominant position of fossil fuels in any modern economy. Although substantial emission reductions can be made through energy saving, increased energy efficiency, and substitution towards less polluting energy sources, the gains are likely only to offset the underlying growth in energy demand, driven by economic growth. Large reductions in energy use can probably be achieved only through major changes in economic structure or lifestyle.

Such changes are politically much more difficult to implement than add-on pollution controls. It has been easier to change the behaviour of a handful of major industries—like utilities—than that of millions of individual decision-makers at the micro level. The former can be achieved by dictates or persuasion from the top; the latter are much less amenable to political intervention.

The choice of political instruments seems to follow from the technical options available. Where end-of-pipe measures are feasible, governments have generally favoured direct environmental regulation, often in the form of technical standards imposed on manufacturers (like the automobile and refining industries) or utilities. This has produced tangible results because technical solutions can be applied effectively by a limited number of large producers of pollution (or producers of polluting 'devices', like cars). Even if such across-the-board regulation has entailed heavy costs, as mentioned above, they have enjoyed popular and political support because they are seen as improving the local and national environment.

Another type of policy instrument frequently applied to the energy sector, partly for environmental purposes, is government-imposed fuel switching. In some countries the authorities can take such action by means of public ownership of major utilities; in others discrete persuasion is preferred. In either case, the government uses its political weight to sway the energy industries to favour certain sources over others for political and not market reasons. Environmental considerations have, so far, not been the primary political motivation, but they have played a role, for example in gas for coal substitution in recent years.

As for economic instruments, we have noted that, despite general expressions of support for such policies, in practice there are few examples of successful implementation in the energy sector—the introduction of unleaded petrol being the major exception here. In applying economic instruments, governments face not only uncertain political reactions, but also the risk of damaging the competitive position of domestic industries. Across Europe, business organizations have warned that additional taxation, whether motivated by environmental or any other concern, can easily lead to loss of markets and jobs. Governments have come close to the limit of effective environmental regulation of the energy sector at the national level. This tendency towards 'exhaustion' of the scope of national-level political intervention for environmental purposes can explain the increasing interest in recent years in international collaboration, in particular within the European Union.

8
The Role of International Policy Bodies

The international character of energy and environmental issues has meant that there has long been a need for international coordination. The treaties that established the European Economic Community in the 1950s had ambitions for far-reaching collaboration in energy production and trade, which came to very little. It was not until the oil crisis in 1973 that Western countries came up with a joint response to an energy problem: the International Energy Agency was set up as an emergency measure. Otherwise national governments remained in control of energy policy. The first serious challenge arose when the European Commission launched its Single Market initiatives in the late 1980s. In this chapter we shall describe the different responses to the main proposals for a more transparent European energy market.

Growing trade across borders also created a need for coordination of environmental policy. We trace here the origins and development of the slowly emerging EU response to these issues. Finally, we shall touch briefly on recent attempts to coordinate energy policy across the East–West divide in Europe.

8.1. EU Energy Policy

Legal Foundations

Two of the three treaties that formed the initial legal basis of the EU directly concern energy. These are the treaties establishing the European Coal and Steel Community (ECSC) from 1951 and the European Atomic Energy Community (Euratom), signed together with the Treaty of Rome in 1957 (the EEC Treaty). The EEC Treaty does not make specific provision for a common energy policy. On the other hand, energy is not exempted from the general rules regulating economic activities.

The fundamental objective of the ECSC Treaty was to secure free trade and a level playing field in the markets for steel and coal within the Community. The emphasis was on steel: coal was included largely because of its importance to the steel industry. Thus, the ECSC established a framework for cooperation on industrial policy for a specific sector, rather than laying down the basis for a common energy policy.

In Euratom, on the other hand, the aim was to cooperate in order to develop a specific energy sector—nuclear energy. Through pooling the resources available in all member countries, technology was to be developed to facilitate rapid growth of the European nuclear industry. However, because of the dominance of US nuclear technology in early phases of European nuclear power programmes and the divergent domestic considerations of EU governments, Euratom never succeeded in establishing a common policy regarding nuclear power. Instead, the nuclear policies of member states followed widely divergent paths. By the early 1990s, policies seemed to be converging, in the sense that all member states except France had adopted effective moratoria on the construction of new nuclear plants. In the UK, no final decision has been taken. What Euratom offers today is common funding for nuclear research and financial support, through the European Investment Bank, for the construction of nuclear power plants. It also has an important role in control of the use of and trade in nuclear materials.

The most important rules concerning energy are contained in the EEC Treaty. Because energy is not explicitly mentioned, the general rules on competition and free trade are in principle applicable to energy services and products as to all other sectors. Articles 30, 34, 57, and 66 of the Treaty regulate the free movement of goods and services. Articles 85, 86, and 90 regulate competition and abuse of market power. These parts of the EEC Treaty are especially interesting in respect of the gas and electricity sectors, where monopolies and trade barriers at the national level are the rule, not the exception. In reality, competition rules have never been enforced concerning the grid-based energy sectors.

Member states have, under certain conditions, the right to grant exclusive supply rights to energy companies, for example the right to supply electricity or gas within an area. This is done to ensure that public-service obligations are fulfilled. Such monopoly rights are, however, allowed to interfere in trade between member states only to the extent necessary for these tasks to be performed, i.e. that they produce a service of general economic interest. On the other hand, internal trade is not to be affected in a way contrary to the interests of the Community.

In the negotiations for the Maastricht Treaty, the EU Commission proposed to include a chapter on energy policy, but deep disagreement on integration policy among member states made this impossible. The main opponent was the UK, which after large-scale privatization and opening of its national gas and electricity supply systems to competition has become the country 'without' an energy policy. The UK has gone further in the direction of liberalization than the Commission has proposed so far and further than any other EU country. The UK opposition was part of its strategy to limit the scope of the supranational authority of the Commission. Energy policy, it was felt, should remain an area where the principle of 'subsidiarity' applied (the subsidiarity principle states that decisions shall be made at the lowest possible level appropriate to the particular issue at stake). However, some elements of the original energy chapter survived in the Maastricht Treaty. Articles 129*b* and 129*c* in the chapter on Trans-European Network state that the EU shall promote the development of energy infrastructure and the opening up of energy transport. EU structural funds will be used for this purpose.

Application of the EU Rules

For oil products, the EU has succeeded in removing trade barriers within the Union. The Treaty has also been used to harmonize oil-product standards, not only with the objective of removing barriers to trade but also as a part of the common environmental policy. The removal of trade barriers for oil products has followed almost the same track as the process for other industrial products.

Treaty rules governing the internal market are, however, far more difficult to apply to electricity and natural gas. Because of the wide differences among member states in terms of the structure and ownership of the electricity and gas sectors, a purely legalistic approach would run the risk of initiating unpredictable processes and creating arbitrary results. It is fairly straightforward to attack legal monopolies, like the import monopoly of EdF in France. Similarly, it can be argued that *de facto* monopolies result from the demarcation contracts in Germany, and as such are an abuse of dominant position and thus illegal. However, the effect of a European Court decision in these two cases would be highly dependent on the structure of the industry and the policy pursued by different member states. A court decision rendering the demarcation contracts illegal could initiate radical changes in Germany, due to the large number of companies and strong opposition in energy-consuming industries to today's monopolies. A decision against EdF's import monopoly would not necessarily affect the company's monopoly position at all. EdF could probably maintain control of its domestic market, and at the same time compete vigorously in the German market. Thus, a pure case-law approach might well affect member states differently, be unpredictable, and not necessarily increase competition.

For this reason, the Commission has so far followed a dual approach: cautiously applying existing competition laws on the energy sector in terms of case-law, and proposing the introduction of new sector-specific regulations through new directives.

The Commission has initiated procedures against several member states which have formal import and export monopolies for gas and electricity in their national legislation. This concerns, for example, France for gas and electricity, Denmark for gas, and the Netherlands and Italy for electricity. Since 1988, the Commission has also adopted a more stringent attitude towards coal: subsidization of national coal production is to be transparent and part of restructuring and rationalization programmes aimed at closing uncompetitive production. Only subsidization which meets these criteria, and therefore is limited in time, will be approved by the Commission. Against this background, the Commission has intervened in the discussions on the future of the German coal policy.

The EEC monopoly rules have been tested against agreements between energy companies—but without arriving at unequivocal results. One example is the SEP case, where the Commission insisted on applying the EEC Treaty's competition rules to the electricity sector. The conflict involved an agreement between SEP, the company responsible for coordinating electricity production in the Netherlands, and distribution companies, granting SEP the exclusive right to import electricity. SEP argued that it needed full control over imports in order to be able to plan new capacity and to guarantee electricity supplies in the Netherlands. The Commission did not accept this. According to the Commission, SEP should be able to satisfy the public-service obligation even if other parties, like distribution companies or industries, imported directly. In a later case also concerning the Netherlands, the Almelo case, the European Court of Justice ruled that an exclusive supply agreement between a production and a distribution company in the electricity sector may not be in violation of the general prohibition of cartels in the EEC Treaty. The limitations on trade following from such an agreement may be acceptable if necessary for the performance of public service obligations. The question of the extent to which public-service obligations may be used for restricting trade is thus left unanswered.

In Germany, the national monopoly control office, Bundeskartellamt, has had recourse to EEC monopoly rules in its endeavours to limit the use of demarcation agreements among the major energy companies and concession contracts for local supplies of gas and electricity. These agreements are allowed by German legislation, but the Bundeskartellamt has challenged them for violating the EEC Treaty. National monopoly legislation is superseded by the EEC rules when internal trade is affected. On this basis, Bundeskartellamt decided in 1993 that the concession contract for electricity supplies to a German municipality bordering the Netherlands, the town of Kleve, was invalid. Also under attack has been the demarcation agreement between two of the major German gas companies, Thyssengas and Ruhrgas, where Thyssengas's reserved supply area according to the agreement borders on the Netherlands, and where the internal trade in gas could be adversely affected by the agreement. No final decisions have been taken in these cases.

EU Energy Cooperation after the First Oil Shock

The oil-price shocks of the 1970s were taken by the Commission as a proof of the need for a common energy policy. There was, however, disagreement among member states on the content of such a policy, due to differences in the structure of energy supply and demand between member countries and the ensuing conflicting interests. The search for a common energy policy in Western Europe has been heavily influenced by the national energy companies, whether privately or publicly owned, and their leverage over the policies adopted by their governments. Member states have preferred to formulate energy policy within their own national institutional frameworks in order to protect their own industrial interests and the interests of their national energy companies. As a result, energy policy has remained the preserve of national governments for the past twenty-five years, despite the continuing rhetoric on a common EU energy policy.

The oil crises did demonstrate that the industrialized countries had at least one thing in common: their dependence on oil. As part of a strategy to deal with this, US President Nixon and Secretary of State Kissinger took the initiative to set up the International Energy Agency (IEA) in 1974, which all EU member countries except France joined. France opposed the creation of IEA mainly because of the dominant role played by the USA. While the rest of Western Europe started to prepare and build up a common response programme, France decided to develop its own foreign-policy response to OPEC and the Middle East, and to avoid steps which could be perceived as a confrontation between oil-consuming industrialized countries and OPEC.

The IEA has autonomous status though it forms part of the OECD. Its Governing Board consists of ministers and senior civil servants from member countries. The secretariat has been granted the authority to act on behalf of member countries in times of crisis. Each member country is obliged to have oil reserves corresponding to ninety days' imports. Furthermore, rules have been established for oil sharing and demand response which the member countries must follow in crises.

EU member states had a common interest in reducing oil imports, but the ensuing energy-policy responses clearly reflected diverging interests among the countries. Strategies took into account the interests and structure of the domestic energy sector. Disparate owner structures and market organization in the energy sector entailed important premises for the way different countries handled the oil crisis. In France, the main response was to expand the state-owned nuclear power production. In Germany the main strategy was to diversify energy supply, promote privately owned nuclear power, and protect national coal production.

One of the few specific energy-policy measures adopted at the EU level was the 1975 decision to prohibit the use of natural gas in new power plants. This directive was decided on a German initiative, copying a domestic policy measure. Gas was considered a premium fuel, whose use in power generation should be restricted, so as to direct its use towards areas with higher efficiency. This directive was never effectively implemented and was abolished in 1989.

The Internal Energy Market

In 1988 the Commission presented a working document on the creation of the internal energy market, with an analysis of barriers to integration. The conclusion was that an internal energy market was needed for the realization of an internal European market. Integration of energy markets and cooperation are seen as offering benefits in terms of lower costs of supplying energy, higher efficiency in energy use, and even improvements in security of supply. Following this, and in response to the Single European Act, the Commission has proposed several new measures. The most important are three measures proposed in 1989:

- Price transparency for gas and electricity
- Transit rights for gas and electricity
- Cooperation on infrastructure and coordination of investments in energy projects of common interest

All three proposals met with strong resistance from the energy industries and from most member countries. After protracted negotiations the price-transparency proposal was accepted. This directive obliges member states to set up uniform price statistics for gas and electricity, an area where differences in tariff principles and statistics make comparisons difficult. The proposal for investment coordination was completely rejected.

The proposal for transit rights for gas and electricity was controversial. The directive guarantees the right of the established gas and electricity companies in any of the EU countries to transit another member country when necessary for their exports or imports of gas and electricity. It was not a proposal for third-party access (TPA), but in the main confirmed the existing practice of cooperation on transit established informally between the main electricity and gas companies engaged in international trade. The companies opposed the transit rules, fearing that they were only the first steps in the direction of a liberalization and the adoption of TPA rules. After long and tough negotiations, the proposal on gas was approved by a majority vote, with several member states voting against. Similar rules for transit of electricity were approved shortly after.

These directives are limited in scope, but in the view of the Commission they constitute the first phase in a liberalization of the natural gas and electricity markets. A second phase consists of opening the grid to TPA. The content of a third phase has not been defined, but the Commission would probably wish to follow the ideas of common carriage in the grids, as indicated in the 1988 working document on the creation of the internal energy market.

The protracted discussions on the Commission's liberalization ideas, from the 1988 internal energy market paper to the adoption in 1996 of a directive on limited TPA for electricity, demonstrate how contentious the issues of liberalization are. Originally, the Commission's liberalization proposal concerned both gas and electricity, but the two sectors were separated in the political process. The conflicts of interest between the member states are strongest in the electricity sector. It was therefore decided to solve the electricity issue first. An attempt to resolve the less contentious gas issue first would probably have been blocked due to fears of prejudging solutions to be found for the electricity sector. The compromise solution for electricity strikes a balance between liberalization and protection of existing companies and also has built in various safeguards, allowing governments to keep on controlling major policy aspects of the electricity sector. Following the adoption, in June 1996, of the directive on the electricity sector a directive for the gas sector is being prepared.

The directive for the electricity sector foresees: a liberalization of the access to generation of electricity; nomination of a system operator; unbundling of the accounts of the system operator and the single buyer; a liberalization of large users' purchases and a minimal opening of the total market; non-discriminatory access to the grid in a system of negotiated TPA, or a single buyer in control of the downstream market. For the access to electricity generation, the directive removes exclusive rights to plant construction, which have existed in some countries. Governments will, however, still have the power to decide on the fuel choice in new capacity. Related to this, member states have the right to impose public-service obligations, such as security and price of supply or environmental protection. These obligations must be 'clearly defined, transparent and verifiable'.

This allows countries to continue long-term planning of the electricity sector, if they wish, and it also allows them to favour fuel and technology choices in electricity generation which are uncompetitive in a liberalized environment. The additional costs caused by such national policies may be levied on all users. This implies that additional costs of national policies will be part of system costs. It is for the Commission to secure that national policies are within acceptable limits for the functioning of the internal market.

Unbundling implies an obligation to separate the accounts of the electricity companies for activities related to generation, transmission, and distribution, in order to secure cost transparency and prevent cross-subsidization. The unbundled accounts should be in a form identical to the accounts of a company who is active in one of these activities only. This means in practice that the information value of the unbundled accounts to outsiders may be limited. In most countries, the published annual accounts contain little information which can be used analytically for the determination of real costs.

The member countries will be obliged to open the market for large single users, i.e. users consuming more than 100 GWh per year. It was not possible to obtain an agreement on an opening of the market for distribution companies' purchases. Distribution companies are not included in the definition of large single users. An important reason for this omission is the French policy of protecting the vertically integrated national company, EdF.

In order to secure—at least in principle—that national markets are opened for competition to a comparable degree, countries are obliged to implement national measures, which from the outset opens 22.5 per cent of the national electricity market for competition. The share of the market to be opened increases to 28.5 per cent after three years and to 32 per cent after a further six-year period. If a country decides only to open for large individual users, the threshold for large users will have to be reduced. When a country decides to include distribution companies in the opening of the market, they may be allowed to reduce the threshold for large individual users less than countries excluding distribution companies from the liberalization.

There is a choice between opting for negotiated TPA or for the so-called single buyer system. Originally a French idea, the single buyer concept was launched to provide a barrier against uncontrolled competition in the electricity and gas market. There are some important differences between the two concepts. In a system of TPA, the eligible consumers (large industries, power plants, and distribution companies) will be allowed to shop around for supplies. The grid owner will be an intermediary, simply providing transportation services. Under a single buyer system, the eligible consumers are allowed to enter into a contract with a competing domestic or foreign supplier, but the electricity should be resold to the single buyer, who remains in control of the market downstream. The single buyer is obliged to purchase the volumes offered at a price which corresponds to his own resale price minus its system costs (transportation and other costs). Under this system, parallel imports become triangular deals involving the foreign supplier, the single buyer, and the eligible consumer. The eligible consumer can make an economic gain on this transaction and becomes a sort of electricity trader entering in between the foreign supplier and the single buyer.

The crucial issue in the single buyer system is the price paid by the single buyer for electricity offered. In this system there is an incentive for the eligible consumer to contract electricity on conditions meeting the needs of the single buyer and not his own needs. If the electricity offered, taking into account all the delivery conditions, is worth less for the single buyer than for the eligible consumer and granted that electricity supply contracts have many other parameters than the kWh-price, the system could easily become less competitive than TPA. In fact, the single buyer concept was put forward by the French Government to protect EdF's and GdF's market position.

In the EU, this has been a point of disagreement between the member states. The directive includes a clause according to which the two approaches to system access must lead to 'equivalent economic results and hence to a directly comparable level of opening up of markets and to a directly comparable degree of access to electricity markets'. The notions 'equivalent economic results', 'comparable level of opening up', and 'comparable degree of access' are not well defined, but are drafted to appease the reservations of the most liberal countries. There is an obvious risk that for example Germany, opting for negotiated TPA, will open its market to more competition than neighbouring France, who will opt for the single buyer system.

There are important question marks concerning the speed of liberalization in the EU and on the national level:

1. *Is the TPA directive only a first step*? The EU liberalization proposal envisages relatively narrow limits on access to the grid and for the eligibility of users. The Commission considers the directive to be only the first step in a liberalization and integration process. It is likely to be followed by proposals

enlarging the scope for competition, but it is questionable when and how the member countries will be willing to accept a new and more far-reaching regime.

2. *Is a partial liberalization stable?* A possible scenario, based on the UK experience, from the Netherlands, and from other countries, is that a limited opening of the market as in the EU directive could be the start of a process, which leads to a further opening. For example, it may be difficult to defend on the political level that in some countries only large users of gas and electricity benefit from TPA, whereas in other countries distribution companies also have access. Another potential cause for instability could be differences between countries in the definition of large eligible users. The cautious opening of the market, as in the EU liberalization proposal, could therefore trigger a more radical liberalization. The speed of such a process is unpredictable.

3. *How strong will the neighbour effects be?* For both gas and electricity, national liberalization policies have impacts on neighbouring countries. Until now, the continental electricity markets have been shielded from such effects. The cable connection to UK is on the continental side controlled by EdF and, between the continental countries the company with the most important export potential is EdF, which has shown no interest in provoking a conflict with neighbouring countries about direct access to the markets. However, liberalization in Denmark has been provoked by market developments in Norway and Sweden. Further ahead, a liberalized Dutch system could have an impact on the German market and the German regulatory system.

8.2. EU Environmental Policy

The Community's first high-level expression of environmental concern came at a meeting of the heads of state and government in 1972 shortly after the UN Conference on the Environment in Stockholm. In 1973 the first action programme for the environment was put forward. The plan presented three basic principles for the development of the environmental policy of the Community: (i) to reduce and avoid pollution by developing protective measures and by introducing the polluter-pays principle; (ii) to develop national as well as Community regulation in order to protect and improve living conditions; (iii) to support international initiatives for dealing with problems not tackled satisfactorily at the national or regional level.

In the period from the early 1970s until today, EU institutions have adopted extensive environmental legislation, including emission regulations, air- and water-quality standards, noise levels, storage and handling of dangerous substances and chemicals, nuclear safety, nature conservation, and general directives concerning information, organization, and monitoring of the environment, as well as financial resources for environmental research.

Legal Basis of Environmental Policy

Environmental policy was not a part of the original treaties of the Community. The problem was resolved by applying Articles 100 and 235 of the Treaty of Rome. Article 100 provides the Council of Ministers with authority to harmonize laws when this promotes the accomplishment of a common market. Article 235 empowers the Council of Ministers to create legislation in policy areas not specifically foreseen in the Treaty. Whereas decisions under Article 100 can be taken by a majority, decisions under Article 235 require unanimity.

The constitutional breakthrough for the development of environmental policy in the EU came with the entry into force of the Single European Act in 1987. This Act has a separate section devoted to the Environment (Part VII, Article 130*r-t*). This meant that environmental policy was now defined at the same level as other policy areas dealt with by the Treaty of Rome. Emphasis is placed on integrating environmental concern with other policy areas. In addition, it is clearly stated that environmental decisions adopted on the basis of the Environmental Article 130 shall not prevent member states from developing more ambitious national environmental policies. Regulations adopted pursuant to the environmental provisions are therefore commonly known as 'minimum directives'.

The Internal Market and Environmental Concerns

The primary objective of the Single European Act was to promote the development of the internal market. The main institutional change was qualified majority voting on issues related to the establishment of the internal market. Much of the subsequent legislation concerning harmonization of standards for traded products has clear environmental implications. These environmental regulations, termed 'maximum directives' are so designated for purposes of preventing member states from developing stricter or less strict environmental regulation than the EU. This has given rise to a number of disputes among member states and between individual member states and the Commission. In some instances member states have applied for exemptions from common EU internal market rules on the basis of environmental concern, referring to the Treaty of Rome Article 100*a*,4 and 36.

Instances of the Commission granting exemption from internal market rules are few and far between. Doing so on a broad scale would undermine the very objective of the internal market. However, since the late 1980s, increasing pressure has been brought to bear on the Commission to take environmental concern into account when drafting new internal market rules. For the environmentally ambitious member states this reduces the urge to apply for exemption from common EU internal/environmental regulations (see Dahl, 1995).

The last major institutional change affecting environmental policy came with the Maastricht Treaty, which opens the door for majority voting on legislation pursuant to Environment Article 130, but with exceptions for, *inter alia*, decisions concerning energy supply. This is regarded as an important step forward for the development of environmental policy, but it also underlines how member states have insisted on maintaining control over the essentials of energy policy.

Implementation of Environmental Policy

Though much of the environmental policy is drawn up by EU institutions, member states have responsibility for its implementation, as is the case for other areas of harmonization. This may often prove problematic. Many of the EU legislative acts have not been properly translated into national legislation and are thus not implemented in the member countries. The procedures against member states for infringement of their obligations are cumbersome and time-consuming.

The establishment of the European Environmental Agency in 1993 may have implications for implementation of environmental regulations. The Agency was created to provide objective information and monitor the implementation and enforcement of EU environmental legislation. However, little hope remains that the monitoring part of its mandate will enter into force.

Environmental Regulation of the Energy Sector

In 1984 the EU took the first significant step towards integrating environmental concern in the energy sector. According to the directive (84/360) for certain industrial plants, new industrial energy-using installations are to be authorized in advance, in order to prevent or reduce air pollution. The directive is based on the concept of imposing the use of best available technology.

The 'Large Combustion Plant' directive (88/609) is one of the most important environmental regulations related to the energy sector in the EU. It limits emissions of sulphur dioxide (SO_2) and nitrogen oxides (NO_x) from new fossil-fuel power stations and other large installations, such as oil refineries and large industrial boilers. The directive also restricts emissions of dust. In the context of policies to combat acid rain, emissions from existing electricity plants are regulated in a 'bubble' concept: total national emission 'ceilings' of SO_2 and NO_x have been imposed with phased and different reductions by member states. Some countries are obliged to reduce their emissions by 60 per cent, while others have been allowed an increase. The allocation of emission reductions partly reflected the differences in the economic development of the countries, but was also the result of a political compromise. Following the revised sulphur protocol signed in Oslo in 1994, many EU member states have agreed to a further (80 per cent) reduction in their SO_2 emissions.

The Community started to develop a policy for climate change in 1988, following an earlier resolution from the European Parliament on the subject. Prior to the Second World Climate Conference in November 1990, the Energy and the Environmental Council agreed to stabilize CO_2 emissions in the Community as a whole by 2000 at 1990 levels, on the assumption that other leading countries committed themselves to similar objectives. This commitment was followed up by a communication to the Council suggesting four directives: an energy-efficiency programme, research to promote alternative sources of energy, a monitoring mechanism for CO_2 and other greenhouse gases, and Union-wide taxation on energy and carbon emissions.

Of these, the former two have been scaled down to modest dimensions as policy recommendations to member states. The proposed carbon tax has created year-long disputes among governments and within EU institutions, without any concrete results, despite intense groundwork from several governments. It has become apparent in this process that it is extremely difficult for member states to agree on policy measures with strong impact on the energy sector at the Union level. This reflects disagreement on the priority of environmental concerns, as well as irreconcilable positions on the issue of EU powers *vis-à-vis* member states. The carbon tax discussion was furthermore complicated by several of the member states' insistence on maintaining taxation policies in the realm of national competence.

The EU stabilization objective remains in place, however, as does a reporting system (the 'monitoring mechanism') to keep track of measures taken by member states to implement their national greenhouse targets. These are regularly reviewed by the Commission in consultation with government representatives.

It follows from the above that the European Union has an important role to play in the formulation of environmental policies. Member states increasingly recognize that it is essential to work towards common standards on issues that affect major industries. With increasing cross-border trade, separate national rules, whether affecting emissions, levies, or taxation, come under growing pressure. However, governments are unwilling to relinquish their environmental ambitions for the sake of EU unity. Consequently, the actual and potential conflicts between the harmonization needed for the creation of the internal market and differences in national environmental priorities are unlikely to disappear in the near future.

8.3. Emerging East–West Energy Cooperation

The European Energy Charter

The process that led to the European Energy Charter was initiated by former Dutch Prime Minister Ruud Lubbers. It was from the outset based on a perception of mutual interests: Western Europe has large and expanding energy markets; Russia has vast natural resources, but lacks the finance and technology necessary to develop its oil and gas reserves. The main objective of the Charter is to create a basis for energy cooperation between the West and the East of Europe, politically as well as economically. The intention is to fuel economic recovery in the East by establishing a free-trade regime for energy products, as well as to establish a framework for the protection of investment and reduce political risk for Western companies investing in the East.

The Charter was signed in The Hague on the 17 December 1991 by forty-eight states, including all the countries of the European Union (as well as the Union itself), all the EFTA countries, nearly all the countries of Eastern Europe and the former USSR (except Turkmenistan), three Mediterranean countries, and the United States, Canada, and Japan. The Charter is a declaration of intentions; after three years of negotiation, it was finalized in September 1994. The final Plenary Session of the European Energy Charter Conference was held in Lisbon in December 1994. Because of difficulties in reaching agreement among all the fifty participants, sector-specific issues are dealt with in protocols attached to the Charter. An Energy Efficiency Protocol was completed in 1994. Protocols for nuclear safety and hydrocarbons have not been finalized.

With its protocols the Charter covers all important aspects of cooperation in the energy sector, and deals specifically with measures to liberalize investments and trade in energy. Cooperation is based on the principles of transparency and non-discrimination, also regarding taxation. Even if many of the states involved in the negotiations are not members of the General Agreement on Tariffs and Trade (GATT), the basic National Treatment and Most

Favoured Nation principles are incorporated in the Charter. These principles mean that investors from any of the signatory countries are not to be treated less favourably than companies of the country in which the investment is made. The principles are activated when the investment is made, not in the pre-investment phase, which will be dealt with in a separate protocol.

Even though the National Treatment and Most Favoured Nation principles are used, the sovereign rights of governments are maintained in the Charter. Signatories recognize the sovereign property rights of states to their energy resources. Each state has the right to decide the geographical areas to be made available for exploration and development of its natural resources. When the negotiation process started, about thirty states were involved, but since then another twenty have joined. This alone created considerable problems during the negotiations, as have economic and institutional difficulties within the former Soviet Union. Many of the countries involved do not have the appropriate legal systems foreseen in the original Charter.

Europe Agreements

In 1991 the EU negotiated the so-called Europe Agreements with the Visegrad countries (Poland, Czechoslovakia, and Hungary) which include:

- Free trade in goods, as a principle
- Rules for migration, establishment of companies, and for trade in services
- Rules for foreign payments, competition, and approximation of legislation
- Economic cooperation
- Financial cooperation

The agreements also state that the objective of the Visegrad countries is full membership in the EU.

The free-trade arrangements are the most important part of these agreements. The target is to create free trade within a period of ten years. Duties and quantitative restrictions are to be abandoned within a year of the signing, but for some sensitive products this is extended to a four- to six-year period. The products subject to transitional rules are clothing, textiles, shoes, iron and steel, furniture, and glass products. For farm products, the EU's agricultural policy establishes special rules which severely limit access to EU markets. On their side, the Visegrad countries are allowed to maintain protection of their markets.

On the surface, the agreements appear liberal in relation to the previous trade barriers. The EU has, however, maintained the right to apply anti-dumping rules against exports from the Visegrad countries, as the Union has frequently done in the past. More than half of the 213 anti-dumping measures applied by the EU in the 1980–6 period were against formerly centrally planned economies. One would expect that anti-dumping in particular will be used against sensitive products such as chemicals, iron, and steel, and that other restrictive measures will be used against agricultural products, as provided for in the Common Agricultural Policy. Such cases have already materialized during the first years of the Europe Agreements, leading to bitter complaints from the Visegrad countries about EU protectionism and negative implications for their economic development.

Apart from the general anti-dumping rules, the Europe Agreements include protective clauses which give wide possibilities for reintroduction of quantitative restrictions and import duties in the case of damage to EU industries. These clauses have been used to reintroduce quantitative restrictions on steel, in response to the increase in steel exports from Central Europe. The restructuring of the Visegrad economies is heavily dependent on access to Western markets. There is a risk that EU protectionism will add to the uncertainties that any foreign investments in Central Europe are subject to, and in this way impede the modernization process. This will also influence the process of reform in the energy sector as well as overall efficiency in energy use.

Both the European Energy Charter negotiations and the bargaining between the EU and the Visegrad countries demonstrate that the critical economic issues in the East–West relationship remain unresolved, despite attempts to establish a stable framework for mutually beneficial interaction. It is still an open question whether the enormous untapped natural riches of the former Soviet Union will be exploited and brought to Western markets. The economic future of Central Europe is equally unpredictable. It could turn inwards in an effort to maintain as much as possible of past structures—or it could open up to integration with the West. The ramifications of this critical choice for energy development, and the environmental future of Europe as a whole, are serious indeed, as we shall see in Part II of this book.

Future Developments
1995–2020

9
Paths for Future Developments

In the first part of this book, we have analysed the evolution of the energy sector in Europe over the past twenty-five years. We have described energy supply and consumption trends and developed a broad understanding of the factors that have shaped the energy sector, its regulatory systems, its corporate structures, and the role of energy and environmental policies. We have seen that the underlying structures, with very few exceptions, have been marked by stability and continuity over the decades. Nevertheless, undercurrents of potential change have been identified, some clearly visible, others hardly discernible in the present landscape. We have demonstrated that the established structures are under stress along two broad dimensions. First, the constraints set up by national borders and national governments are challenged by market pressures for international trade and competition among suppliers across Europe. Second, the environmental issues of the 1990s, spearheaded by climate change, could demand a radical restructuring of energy use and production, regardless of political set-up.

Part II is devoted to a study of how these challenges can shape energy and environmental developments in Europe over the next twenty-five years. We start from the insights gained in Part I and the analytical framework presented in Chapter 1. Our analysis will first focus on two broad alternative scenarios for future developments in the political and economic environment in Europe. These depict distinct paths in energy-sector developments, one towards an open competitive European-wide market regulated at the EU level ('liberalization and trade'), the other ('national rebound') within the confines of nation-states. We demonstrate on the basis of the structural pattern analysed in Part I how the energy sector is likely to develop in each of these cases, and summarize by offering a systematic comparison of them.

Then, we proceed to our ultimate question: if environmental concern rises to new heights, how will this affect energy developments over the next twenty-five years? This depends in a critical sense on the direction of development of the energy sector as such, as the options open to policy-makers vary significantly in the two scenarios. The purpose of our concluding chapter is to show the different ways in which a new environmental force will work through European energy structures depending on the overall context defined by the larger political and economic picture.

9.1. Introduction

The principal features of the scenarios are determined by the five forces, described in Chapter 1 and shown in Fig. 9.1, that together set the agenda for the future of the energy sector in Europe. The forces act differently in the two scenarios, giving rise to different energy-sector developments and environmental trends. Three of these forces—political and economic integration,

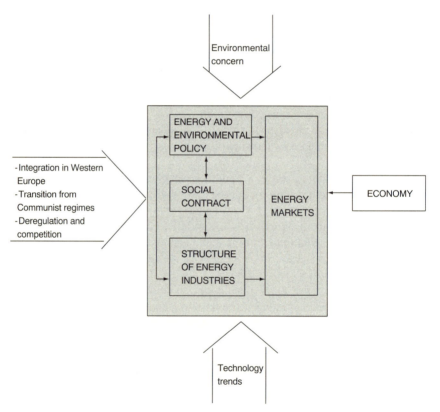

FIG. 9.1. Framework of analysis

the transition from Communist regimes, and the trend towards deregulation and competition—are closely inter-related. We shall consider them as a single cluster of independent variables which can work in either of two directions: towards an open all-European economic system regulated at the EU level, based on increasing competition across borders, i.e. *liberalization and trade*; or, alternatively, a European economy divided into an Eastern and a Western part where national governments remain in charge of fundamental decisions concerning energy supplies and environmental policies, i.e. *national rebound*.

Our fourth driving force, technology, is assumed to be strongly influenced by the overall trends set by this cluster. If the general political/economic development takes the liberalization track, there will be strong commercial pressures to develop technological options that facilitate more open markets with freer access for small-scale actors, possibly down to the household level. Innovations in information technology will be stimulated for this purpose, as well as small-scale renewables.

If the overall trend goes in the opposite direction, the demands for techno-logical development will be very different. The large-scale bias of the dom-inating energy industries will set the agenda for research and development, which will preclude small-scale options and favour solutions that fit current structures, such as clean-coal technologies and improvements in nuclear power.

We shall argue that the fifth force, environmental concern, is independent of the other four. This means that different environmental priorities are compa-tible with both of the scenarios. Each of them therefore includes two environ-mental futures: one where environmental concern and level of political priority remain at current levels, and the other where both rise significantly, forming a third environmental wave by the turn of the century.

9.2. Energy Futures

National Rebound

In this scenario, EU integration is stalled, and Central Europe remains a separate economic system, still influenced by its Communist heritage. In both East and West, national governments remain in charge of all important elements of economic and social policy. Trade in electricity and gas remains under national control, exercised either by state-owned companies or in close understanding between the private and public sector. Governments use their control over the energy sector to promote extraneous objectives, like industrial and regional development and employment.

Energy industries remain for the most part in public hands, as a guarantee of political control. Vertical and horizontal concentration is pursued in order to strengthen national energy enterprises, technically and financially. R&D programmes are tailored to the needs of the large energy industries, and partly financed by the state. Priority is given to coal combustion technologies and large-scale nuclear research. The essence of the social contract inherited from the post-war period remains intact: the state retains control over energy industries, including overt and covert interference in business operations and commercial planning. In exchange, the companies are largely protected from competition in their line of business.

Liberalization and Trade

In this scenario political and economic integration in Europe, including the Visegrad countries, proceeds rapidly in depth and breadth. Political reforms are imposed to ensure free trade in energy, gradually leading to price convergence towards a uniform level. Energy supply industries accommodate to a competitive, transnational business environment, learning to operate as commercial entities, separate from political interference. The greatest remaining political concern (apart from the environment)—energy security—is elevated from the national to the EU level. The top political objective is to stimulate energy markets and to establish appropriate regulatory systems for this purpose, especially for grid-based energy supplies. The European Commission plays a key role in developing and enforcing transparency and competition in this field.

Once such regulation is in place, private investors gradually take more interest in formerly state-owned energy companies. This in turn leads to ownership concentration and conflicts with regulators preoccupied with competition in the market place. In R&D, public programmes are scaled down, with the exception of EU-funded research. In both private and EU programmes the emphasis is increasingly on 'smart' IT that promotes trade and competition at the micro level, as well as small-scale renewable alternatives.

In short, in this scenario politicians are primarily concerned with economic efficiency in the energy sector. They are therefore willing to grant industry a wide measure of independence, provided it can deliver energy safely and at a reasonable cost to consumers and society at large. This 'hands-off' attitude is the crux of the new social contract: the state no longer interferes directly in the energy sector, since it has given up promoting objectives like regional devel-

Table 9.1. *Comparing the two scenarios: an overview*

	Liberalization and Trade	National Rebound
Political and economic integration	Leap forward	Stagnation
Internal energy market	Implemented	Stalled
Government–industry links	Looser	Closer
Energy security policy	EU concern	National concern
Central European links to the West	Integration in EU	Separate economic system
Regulatory system	Transparent, enforced at EU level	National control, industry self -regulation
Ownership	Privatization and concentration	Primarily public, vertical and horizontal concentration

opment and employment which it did so forcefully in the quarter-century before 1995.

9.3. Environmental Futures

As mentioned, we assume that both high and moderate environmental concerns are compatible with both scenarios, since there is no logical link between the ecological dimension and the overall political/economic trends in the two cases. Environmental concern may increase (or stagnate) whatever happens with European integration, but the framework conditions for environmental policies will be radically different depending on the larger context, as we shall see below. First, we need to define the difference between the two values on the environmental scale: moderate (i.e. current levels) and high.

Moderate Concern

In this case, the focus remains on local environmental problems. Air pollution from stationary energy combustion declines as modern pollution-control technologies are mandated and stricter standards are imposed on oil products. In Western Europe the major unresolved local problems are related to air pollution from vehicles and siting of power plants and transmission lines. In Central

Europe there are in addition major environmental concerns related to coal mining, in particular near lignite mining sites.

Regional environmental problems, primarily emissions of acid rain precursors, are addressed in accordance with the objectives established in protocols on SO_2 (signed in 1994) and on NO_x (into force in 1996). Global challenges associated with CO_2 emissions are not addressed through any concerted actions among European countries, nor is there any agreement to act at a broader international level. Measures imposed at national levels are, as before 1995, moderate in scale. They have only minimal effects on energy markets and no impact on energy industry structures.

High Concern

Public concern and political commitment to combat local environmental problems increase compared with the priorities given to these issues in the first half of the 1990s. The implications are strongest for the transport sector. In the case of Central Europe, restrictions are also imposed on lignite mining, as a result of renewed concern for preservation of the natural environment. This also has implications for siting of power plants and large transmission systems for electricity and natural gas. Acid rain and transboundary pollution that contributes to tropospheric ozone causes increasing concern, leading to a strengthening of the protocols on SO_2 and NO_x. The new targets call in particular for radical measures against NO_x emissions. Global warming becomes the dominant environmental issue. European countries make commitments, as part of internationally negotiated targets, to radical reductions in emissions of CO_2 and other greenhouse gases.

In the following chapters we shall analyse how policies and energy-sector structures of the two scenarios impact on energy-market trends (Chapters 10 and 11) and environmental indicators (Chapter 12). Subsequently we discuss the scope for and effects of policies in the two scenarios when there is a rise from moderate to high concern for the environment (Chapter 13). Trends in energy consumption and environmental indicators are based on model simulations by an energy supply-and-demand model—ECON-ENERGY. Detailed results from the model simulations are presented in the Annex. A brief description of the model is provided in Annex B.

10
National Rebound Scenario

In this scenario, economic and political integration in Europe comes to a halt, and nation-states generally reinforce their control over social and economic policies, including the energy sector. Economic restructuring is slow, with continued protection and support to industries that traditionally have been shielded from competitive pressures. Trade increases slowly, with only modest growth in transport requirements. Energy-sector regulation and institutions are amended to meet new economic and political circumstances, but in general developments in energy-industry structures are characterized by consolidation and preservation of established vested interests within the sector. Most countries pursue an active, in some countries interventionist, energy policy. Energy security, industrial policy, and employment considerations are the main priorities directing energy policy. Protection and strengthening of national energy enterprises with tight state control are important means to achieve these ends. Trade in energy between countries is generally regulated and end-users have very limited choice between suppliers of electricity and gas.

The direction of energy policy has a major impact on energy-market developments. The composition of energy consumption and emissions of CO_2 are therefore, as in the past, heavily influenced by policy. However, energy policy is not the only determinant of energy consumption: economic activity and technological options not directly affected by energy policy also play a role.

10.1. Main Features

Integration on Hold

Denmark's 'No' to the Maastricht Treaty in the 1992 referendum marked the beginning of growing public and political resistance throughout Europe to further political and economic integration. There is a perception among key member governments of the EU that their interests are best served by maintaining firm national control over economic and social policies. The globalization of the world economy and technological progress that threatens the living standards of certain social groups are seen as a political challenge best dealt with at the national level. The lack of effectiveness, flexibility, and democratic attributes in the EU force member governments to deal with the complex set of political challenges that arise in the second part of the 1990s. The rules and regulations established through the Single Market process of the late 1980s and early 1990s are not altered, but the ability of the EU institutions to enforce them is rather weak.

European-wide competition in agriculture and selected industrial sectors threatens employment, and as such is a potential threat to social stability in the West. Structural reforms in Central Europe proceed slowly for similar reasons, with market protection for important manufacturing and energy industries remaining in place. Economic and political cooperation between EU and the Visegrad countries is based on the Europe Agreements. In the agreements exemptions, from competition and trade for 'sensitive products', limit trade to what is socially acceptable. Moreover, national regulations also inhibit the free flow of capital.

EU versus National Energy Policies

The individual nation-states remain in control of energy-sector policies. As in the past, these vary between countries, as do energy-industry structures and regulatory systems. There is a common understanding within the EU that the technical characteristics and strategic importance of the energy sector merit different approaches to sector regulation and policies.

Prices and costs of energy supplies continue to show major differences between countries. Social and industrial policy considerations outweigh 'market pressure' for alignment in energy prices. In order to protect employment and permit an offensive industrial policy, energy-intensive industries are offered favourable contracts for energy deliveries that encourage investment in these industries and strengthen their position in international competitive markets. Such contracts are not seen as violating the EU policy on competition and state-aid issues.

Nor is state support to energy industries hindered by EU institutions. Aid to coal production in Germany and Spain remains high. Nuclear power plants increasingly need financial transfers to maintain regularity and safety in production. National security of energy supplies is an important motivation for support both for coal and nuclear, but with coal the most important political considerations are social.

Energy Security

Diversification of energy supplies increasingly becomes a preoccupation of energy authorities. There is no consent to deal with the issue effectively at the EU level. In line with the traditions of the 1970s and 1980s, therefore, member governments use their control over investments in the power sector as the principal instrument for diversification of national energy supplies.

The prevailing structure in power generation, with little competition and the continued progress in the economic attractiveness of 'advanced coal' technologies, reinforces a renaissance for coal, as dependence on natural gas increases. Some of the major European electricity enterprises continue their nuclear programmes; the construction of new plants is actively supported by energy authorities in preference to the less attractive alternatives of imported natural gas or coal.

In Central Europe, energy security still tops the political agenda. Slow progress in integration of energy markets with Western neighbours and obstacles to commercial and technical cooperation with Western energy enterprises make the Central European countries vulnerable to potential supply disruptions. In addition to energy security, social considerations explain the importance of continued national support to coal production.

Stable Industry Structures

Enterprises are granted substantial operational autonomy from governments, providing that they take into account political priorities concerning employment, industrial policy, and the environment. However, there is active governmental involvement in major strategic issues for national energy-sector developments, in particular in securing national ownership in vital companies and in securing energy supplies. Foreign investments in energy-sector activities are relatively modest.

Restructuring and energy-sector reforms are still undertaken to some extent, due to the high costs in energy supplies which represent a burden on energy consumers and public budgets. Uneconomic coal production and overstaffed and inefficient energy utilities represent challenges that are tackled without major alterations to energy-industry structures or supply patterns.

Some energy enterprises attempt to improve their financial position by seeking markets in other countries, which helps prevent a radical down-scaling of activities when growth in domestic markets decelerates. This development is supported and encouraged by governments, as it strengthens the major national (often government-owned) energy enterprises. Hence it leads to horizontal, in some cases vertical, integration of energy-supply industries at the national level.

Technological Trends

Ambitious R&D programmes are initiated in support of the policy goals for performance in the energy sector, i.e. energy security, environmental protection, and cost reduction. The programmes are adapted to the requirements of the energy-supply industry. A major part of R&D resources is directed towards the power sector, in support of the dominant technology and fuel choices—whether coal, nuclear, or natural gas.

R&D funding is also allocated to energy-conservation efforts. Technical standards and rebate programmes are in wide use in order to encourage the adoption of new technologies. In countries with centralized, vertically integrated power systems, 'integrated resource planning' plays an important role in disseminating new end-use technologies.

10.2. Main Trends in Energy Consumption

Central versus Western Europe

For the entire period from 1995 to 2020 total energy consumption increases by 24 per cent, against a total growth of 33 per cent for the quarter-century preceding 1995. This deceleration takes place despite higher economic growth for Europe as a whole. Hence the decline in energy intensity (energy consumption per unit of GDP) gains further momentum after 1995. For both Western and Central Europe, the annual growth in energy consumption from 1995 to 2020 is about half the growth in GDP. However, there are some important variations over time and between regions.

The most striking feature of energy consumption developments is the declining trend in growth rates in Western Europe and the rising growth rates in Central Europe (see Fig. 10.1). This tendency holds for all countries within the two regions.

The lower energy consumption growth in Western Europe is primarily caused by two factors. First, oil and natural gas prices remain stable until 2005 and only start to increase significantly after 2010. Second, saturation in

Box 10.1 *Key assumptions*

Growth in Gross Domestic Product (GDP) differs markedly between Western and Central European countries. In Central Europe the collapse of Communism unleashes a major potential for economic growth. Hence, after the first four to five difficult years of economic recession and restructuring after 1988–9, the economies of Central Europe start to grow at a high annual rate despite prevailing trade barriers between Western and Central Europe. During this period major productivity gains are achieved through transfer of resources from low-value to high-value production, and by the adoption of modern technology. Growth slows down somewhat after 2005, but continued high investments contribute to an economic growth significantly higher than in Western Europe.

	Average annual growth rates, %		
	1970–1995	1995–2005	2005–2020
Gross Domestic Product (GDP)			
Western Europe	2.6	2.6	2.6
Central Europe	−1.0	4.5	4.3
Population			
Western Europe	0.4	0.2	0.2
Central Europe	0.6	0.3	0.2

In Western Europe economic growth rates remain at 2.6 per cent per annum throughout the entire period, the same as from 1970 to 1995. There are differences in growth rates between the high- and middle-income countries of Western Europe, with the latter group having an annual growth of 3–4 per cent per annum.

some areas of energy use, such as space heating and transport, cause a decline in energy consumption growth. Industrial restructuring after the turn of the century also has an effect on energy consumption trends. The opposite trend in growth rates in Central Europe is explained by economic restructuring and continued high economic growth. Industrial restructuring, with the closure of old and highly inefficient industrial plants, is the principal reason for the slow growth in energy consumption before the year 2000. A slow but gradual increase in energy end-use prices also acts to curb demand.

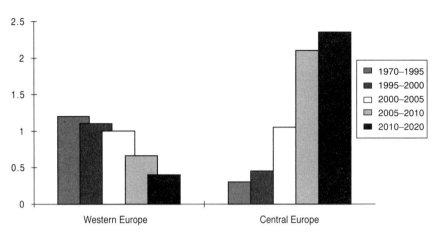

F<small>IG</small>. 10.1. Total energy consumption in Western and Central Europe, 1995–2020 (Average annual growth rates)

Box 10.2 *Energy prices*

Coal prices

International coal prices have declined continuously in real terms since the peak year 1982. The downward trend continues after 1995. Sliding international coal prices are the result of the emergence of an increasingly competitive international market for steam coal. Infrastructure for the coal trade has grown rapidly as coal-exporting countries like South Africa, Australia, Canada, and Colombia have increased their production capacity. At the same time, productivity in coal mining has increased and international coal prices have fallen apace. The trend in international coal prices continues to be determined by the development in mining productivity and transportation costs. Given the vast coal reserves and continued progress in mining technologies, this means persistent increases in productivity and lower coal prices. Given the dispersion of coal reserves, this trend is not disrupted by cartel-like behaviour on the part of producers.

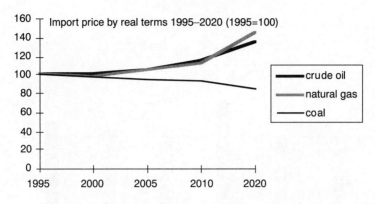

Oil prices

Crude oil prices remain relatively low until the turn of the century, due to continued spare production capacity in the Middle East. In the five years from 1995 global oil consumption increases by almost 2 million barrels per day (on an annual basis). About 80 per cent of the additional demand is met by supplies from the Middle East. Hence idle production capacity in the region is reduced and the occurrence of 'overproduction' becomes less common. This results in an upward trend in crude oil prices, though prices continue to fluctuate, as in the past.

Natural-gas prices

Import prices for natural gas continue to be affected by price paths for competing petroleum products. But as the natural-gas market gradually tightens, prices are increasingly determined by the costs of bringing additional quantities of natural gas to the European market. In 2020 marginal supplies cost 35 per cent more than the average import value in 1995.

Sectoral Development

Developments for the three main end-use sectors are quite diverse; see Fig. 10.2. Transport-sector consumption shows substantial increases in Central Europe, whereas the growth in Western Europe is moderate. Industrial energy use increases only moderately in both Central and Western Europe. In other sectors, notably residential and service sectors, energy use shows fairly strong growth in both regions.

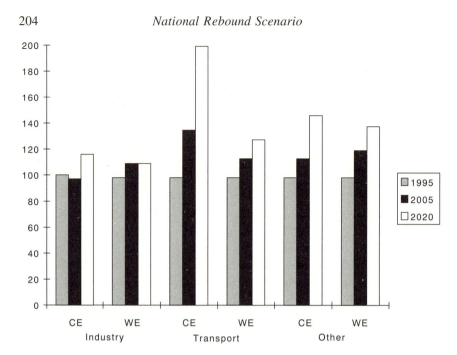

Fig. 10.2. Energy consumption in Western Europe (WE) and Central Europe (CE), 1995, 2005, 2020 (1995=100)

Manufacturing Industry

A stagnation in industrial energy consumption from 1995 to 2005 takes place despite an increase in industrial production of 1.5 and 3 per cent per annum in Western and Central Europe, respectively. After 2005, energy demand remains flat in Western Europe, but increases in Central Europe.

In Western Europe, the average annual decline in industrial energy intensity from 1995 to 2005 is twice as high as during the first part of the 1990s (see Fig. 10.3) but substantially less than in the period from 1973 to 1990, because energy prices stay low. International energy prices remain depressed until about 2010 and energy-intensive industries continue to enjoy advantageous electricity tariffs in some countries. Therefore, investments in energy efficiency and structural changes within industry are not as high as in the 1970s and 1980s. During that period at least one-third of the reduction in industrial energy intensity was the result of structural changes; from 1995 to 2005 structural changes contribute very little.

After 2005 the annual rate of reduction in energy intensity becomes more

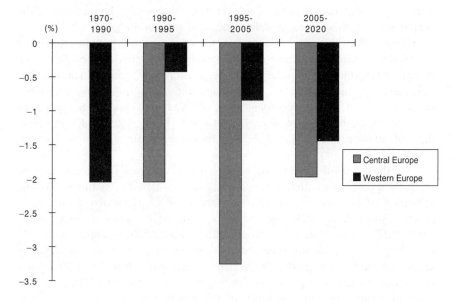

F<small>IG</small>. 10.3. Changes in industrial energy intensities in Western and Central Europe, 1995–2020 (average annual growth rates)

pronounced, at 1.4 per cent per annum. This reduction is a result of higher international energy prices and higher electricity prices, the latter due to a reduction in the surplus capacity of base-load electricity that had prevailed since the early 1980s. Then growth in electricity demand dropped but new generating capacity continued to come on-stream in the same magnitudes as before. As a consequence, earlier long-term contracts for energy-intensive industries are not always renewed on the same advantageous terms as in the past. This, together with enhanced global competition in the manufacturing of iron and steel, non-ferrous metals, and chemical products, imposes a gradual restructuring of West European industry.

In Central Europe the reduction in industrial energy intensity is larger, at more than 3 per cent a year from 1995 to 2005. But from 2005 to 2020 it falls to 2 per cent per annum. This may be considered a major reduction in energy intensity. However, in view of the pervasive inefficiency prevailing in Central European industrial energy consumption until now (see Chapter 2), the decline is in fact relatively modest. There are several reasons for this.

The influx of foreign capital and know-how is limited as trade barriers remain in place, and governments are slow to establish a modern legal framework comparable to that of the West. Consequently, the turnover of the capital stock, technological development, and structural change within industry remain relatively slow, in consideration of the overall growth rate of the economies. Central European governments continue to provide subsidies to the manufacturing industry and the energy sector, for social and political reasons. Consequently, the closure of unprofitable industries in the energy sector (mining, district heating companies) is delayed and certain energy prices (e.g. electricity, heat, gas, and coal) continue to be regulated and kept below the costs of supply. This is also the case for other industries, such as steel and chemicals.

Nevertheless, economies grow rather rapidly, spurring investment in new production processes with higher energy efficiency than the existing stock. The scrapping of obsolete capital left over from the Communist era continues. Despite sluggish restructuring, therefore, industrial energy intensity drops substantially in absolute terms from 1995 to 2005. Subsequently, as the old technologies are gradually replaced and energy prices approach the level of supply costs, further reductions in energy intensities slow down. Industrial production continues to grow at 3–3.5 per cent per annum after 2005, whereas energy consumption grows at a rate of 1–1.5 per cent.

Transport

Developments in the transport sector are influenced primarily by disposable income and volumes of trade flows. Whereas in Western Europe certain modes of transport are reaching saturation, energy demand for transport in Central Europe is highly sensitive to economic activity and personal income. The economic transition in Central Europe, with solid growth in personal income and transport requirements, makes this sector the principal source of growth in energy demand. Transport fuel demand grows on average about 3 per cent per annum. Private car ownership increases substantially but as personal income remains below the average of Western Europe, smaller cars have a greater share in the fleet of Central European countries. Consequently, the average efficiency of the car fleet and the average kilometrage is below the levels of Western Europe.

The downward trend in the growth in transport fuel demand in Western Europe was evident from the early 1990s, when consumption fell to 2 per cent per annum after a growth of more than 4 per cent per annum during the 1970s and 1980s. The growth rate slides to 1 per cent for the period 1995 to 2005 and to 0.8 per cent per annum thereafter. This trend is explained partly by saturation in car ownership in some market segments and because kilometrage per car increases only modestly. Moderate investment in transport infrastructure curbs growth in driving, although this leads to more congestion, which in turn has a detrimental effect on fuel efficiency. On balance, however, fuel efficiency improves slightly despite the increasing preponderance of ownership of larger and more powerful cars as income level rises.

Energy demand in other sectors is concentrated in services and households. The first experiences a strong growth in energy demand, while saturation in Western Europe and energy-efficiency improvements in Central Europe help to contain growth in the household sector.

Service Sectors

In Western Europe, service-sector energy consumption continues to increase quite rapidly, as it did during the 1970s and 1980s. This reflects a strong growth in economic activity within services and limited scope for reductions in energy intensity. However, there is gradually a decoupling between the value of production in service sectors and the need for additional office space, which is the principal determinant for the sectors' energy demand. New telecommunication technologies and office equipment allow companies to reduce their office space requirements. Moreover, companies increasingly introduce flexible working hours and encourage people to work from a home base, in efforts to cut overhead costs. In consequence, there is only a slow increase in the energy demand for space conditioning (heating and cooling), whereas electricity demand rises substantially.

The service sector in Central Europe is still in its infancy in the mid-1990s. The region embarks on a process of radical reorientation and expansion of commercial and public service sectors. Strong economic growth leads to a substantial demand for office buildings, and thereby energy demand for space conditioning and office appliances. In addition, increasing disposable income and consumer spending require the establishment of an acceptable level of commercial services.

Residential Energy Use

Developments in the residential sector are driven by changes in lifestyle and disposable income. Consumption in the residential sector is also determined by the development of new information and communication technologies. Furthermore, people tend to enhance the level of comfort within their home, leading to a further penetration and increased use of residential appliances. Reduction in working hours, flexible working hours and lifestyle changes, including working from home, require increased utilization of living space, which leads to an additional demand for electricity.

In Central Europe these changes give rise to significant changes in the composition of residential energy demand. Energy demand for space heating in Central Europe is decreasing, given the large potential for efficiency improvements of the building shell: economic growth fosters new construction and refurbishment of housing. However, this process is hampered by the slow pace of improvements in the technical and economic performance of district heating. From about 2010 the major part of the efficiency potential from refurbishment of buildings is exhausted, thus energy demand for space heating again points upwards as the demand for housing continues to increase in response to higher income. A rapid increase in the ownership of electrical appliances gives high growth rates for electricity throughout the entire period from 1995 to 2020.

Electricity demand is also the driving force in residential energy consumption in Western Europe. Consumption for space heating and cooking increase only moderately, due to saturation and very low population growth.

The Electricity Sector

For Europe as a whole, electricity consumption increases from 2300 TWh in 1995 to 3190 TWh by 2020. As indicated above, electricity gains market shares in all major end-use sectors. Industrial processes increasingly require electricity for mechanical work and for process heat. The substitution of electricity for direct fuel use is particularly pronounced in Central Europe through the replacement of old machinery and equipment, although this effect is initially offset by major improvements in end-use efficiency.

Despite its increased competitiveness, there is a trend towards lower growth rates for electricity. In the past in Western Europe there has been close to parity between electricity demand growth and economic growth, and in periods an even higher growth in electricity consumption, but from 1995 and onwards a clear decoupling of the two. Between 1995 and 2020 electricity consumption in Western Europe increases by 1.2 per cent per annum, against an average annual growth rate of 3 per cent from 1970 to 1995. This lower growth rate is caused primarily by gradual saturation in the residential sector and efficiency improvements in industry.

In Central Europe electricity grows by less than 1 per cent per annum from 1995 to 2005 as major efficiency gains are exploited, rising to 2.5–3.0 per cent after 2010. This can still be considered low compared to other countries with rapid economic growth. The norm is rather that rapid economic growth is accompanied by even more rapid growth in electricity consumption. The situation is different in Central Europe because electricity intensity (electricity consumption per unit of GDP) is extremely high, and electricity consumption

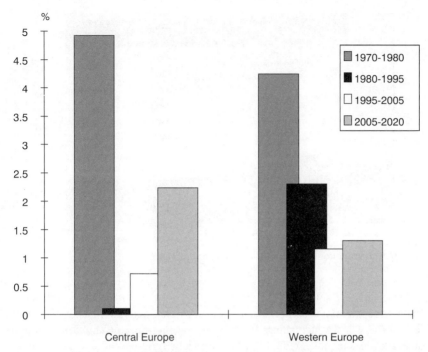

FIG. 10.4. Electricity consumption in Western and Central Europe, 1970–2020 (average annual growth rates)

per capita is also high in relation to the modest income level. This reflects a major increase in electricity consumption during the 1970s and part of the 1980s, before the political and economic collapse.

Energy Consumption by Fuel

Coal maintains a relatively high market share despite the economic and environmental attractions of natural gas. There is a notable reduction in direct coal use, especially in the residential sector, but coal continues to be extensively used in industry and in district heating systems in Central Europe.

The share of oil declines slowly from 42 per cent in 1995 to 39 per cent by 2020. The shift from fuel oils to transport fuel continues, but with less momentum after 2010 as the growth in transport fuel consumption declines. From 2010 fuel-oil demand rises again, after more than thirty years of decline

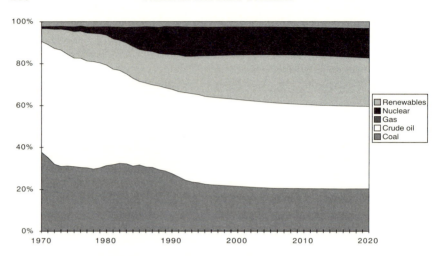

Fɪɢ. 10.5. Energy consumption by primary fuel in Europe

and stagnation. Fuel-oil prices are becoming more competitive in both industrial and residential fuel markets and for electricity generation, as import prices
for natural gas are increasing. Restrictions on fuel-oil consumption for environmental reasons are gradually relaxed as new refining and end-use combustion technologies contribute to reduced air pollution.

Natural-gas consumption increases by 50 per cent from 1995 to 2020. Its
share of primary energy demand continues to increase until 2005, and then
reaches a market share of 24 per cent, ten percentage points higher than in the
early 1980s. Natural gas gains markets in the power sector, industry, and in
households and services. The most notable increase is within industry and
service sectors, where natural gas partly replaces coal and fuel oil. After 2005
natural gas meets increasing competition from coal because of lower relative
coal prices (see Box 10.2) and as advanced coal combustion technologies
become commercial. The market share for both natural gas and coal therefore
remains stable from 2005 to 2020.

Nuclear power plants continue to come into operation; by 2020 annual
production is 1050 TWh, against 830 TWh in 1995. The nuclear share in
primary energy supplies remains at about 14 per cent throughout the period.

Renewable energy grows apace with total energy consumption, i.e. its share
of electricity and heat generation remains 5 per cent. In 1995 some 2 per cent
of this category is hydroelectric power, but only a small part of the growth in
renewables to 2020 is hydro. The remainder comes from the so-called new

renewable energy sources. Electricity production from these sources increases from 30 TWh in 1995 to 230 TWh in 2020, of which biomass and municipal waste are the most important. Their penetration into the market is partly explained by lower costs due to technological improvements, and higher prices for natural gas and oil, but an active policy to promote renewables is also important.

10.3. Energy Policy: Security of Supply

Import Dependence

As described in Part I, energy policy had a significant influence on the composition of energy consumption from 1970 to 1995, in particular from 1973 to 1985 (see Fig. 10.5). In this scenario, governments retain their authority to control the development of the primary fuel mix. The main political objective is, as in the past, to reduce vulnerability to potential energy-supply disruptions and protect domestic energy producers. Despite the stability in fuel composition achieved through active policy measures there is a gradual increase in the import share. This increase started in 1995, and by 2020 reaches the same level as in the mid-1970s.

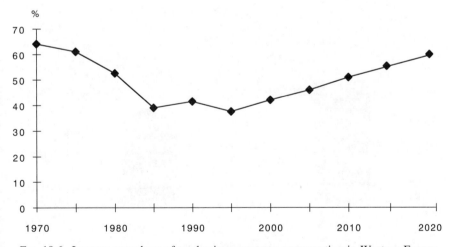

FIG. 10.6. Imports as a share of total primary energy consumption in Western Europe

Oil Imports

The highest share of import reliance and largest increase in import share is for oil. From 1995 to 2020 the import share increases by twenty percentage points to 74 per cent. What causes political concern is not the increasing dependence on non-European oil in isolation, but rather the increasing global reliance on oil supplies from the politically unstable Middle East. In 1995 some 30 per cent of global oil consumption originated from that region. By 2010 this figure has increased to 45 per cent and by 2020 to 60 per cent. Europe's contribution to this development is only modest. It is rather the result of rapid growth in oil consumption in developing countries. Oil provinces outside the Middle East are only able to meet a small fraction of the incremental demand. Therefore any disruption to the oil market will have global ramifications, and reliance on this one commodity in the case of a supply disruption may cause major economic and social damages.

Natural-gas Supplies

Natural-gas markets by contrast are regional in character. The policy issues related to import dependence are therefore somewhat different. There are no major changes in the supply pattern for natural gas from sources outside Europe. Russia remains the dominant supplier, with European imports increasing from 83 mtoe in 1995 to 180 mtoe by 2020. The gas is supplied through

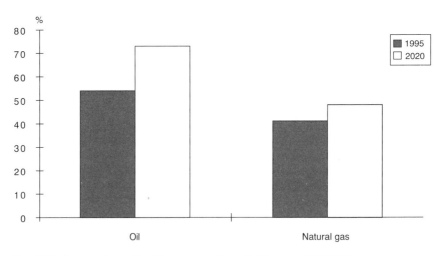

FIG. 10.7. Import shares for oil and natural gas in Europe, 1995–2020

two channels: the 'Brotherhood' pipeline transiting the Ukraine and a new pipeline through Belarus and Poland to the West European market. Algeria, the other major exporter to the European market, increases deliveries from 42 mtoe in 1995 to 100 mtoe in 2020. Most of this gas is landed in Italy through the Transmed pipeline system. In addition a pipeline through Morocco to Spain (Europe–Maghreb Pipeline) transports some 8 mtoe/year from 1997 and with further expansions after the turn of the century. There is also an increase in the imports of LNG (liquefied natural gas) from Algeria, Libya, and Nigeria. New LNG deliveries from these countries are by 2010 on a cost basis competitive with additional production from Norway or other European sources (see Box 10.2). However, since imports from these sources hardly represent any enhanced security of supply, they grow only slowly until 2020.

The increasing imports of natural gas from Russia and Algeria serve as the principal rationale for policy actions aimed at influencing the fuel choices of European market actors. Although concern about high dependence on imports from potentially unstable areas is not new, soon after the turn of century it becomes the principal energy-policy issue, attended by greater political concern than during the 1980s. It also has an important commercial dimension. Commercial actors, both the gas industry and gas consumers, consider natural gas supplies as potentially unstable. This leads to some reluctance in investing in gas infrastructure and capital equipment which is fuelled solely by gas.

10.4. Energy Supplies: The Impact of Policies

Coal Supplies

European countries have vast reserves of coal, which is considerably more expensive to mine than in exporting countries like Australia, the United States, and South Africa. All the same, European countries continue to produce coal at the 1995 level for the entire period up until 2020. In Central Europe, production of lignite for power production actually increases after having declined during the years of economic recession and restructuring after 1990.

It is important to distinguish between hard-coal and lignite production. Hard-coal production tends to require state support in order to compete with imported coal and other fuels. Lignite, though inferior to hard coal in heat content and environmental qualities, is for a larger part economically viable, if used close to the mining site. However, its attractiveness is very sensitive to environmental regulations, not least concerning waste disposal and land-use with open-cast mining.

State protection of the coal industry continues to be financed through price support and direct financial transfers from state budgets. The latter are largely earmarked for restructuring and modernization of coal mines. There is therefore a substantial improvement in coal-mining productivity in Central Europe, as well as some continued improvement in Western Europe.

Hard-coal Production

Poland, Europe's principal producer of hard coal, has a production level of 62 mtoe in 1995. By the turn of the century production drops slightly to 60 mtoe. Beyond the year 2000 production maintains its slow decline, despite major efforts by the government to keep the outlets for Polish coal. Both domestic and foreign demand decline. On the domestic market, energy-efficiency improvements in industry, a structural decline in residential coal demand, and sluggish demand growth from the power sector contribute to the depressed market. In addition, the relatively slow modernization of the coal sector means that the productivity gains and cost reductions are inadequate to make exports profitable. Polish coal cannot compete on the international market with seaborne supplies. The best prospects are in neighbouring countries within a radius of 500 km (Radetzki, 1994*a*). Part of Germany, the Czech and the Slovak Republics, and Hungary fall within this area. Exports to these countries increase, but are significantly hampered by import regulations. From 1995 to 2005, therefore, Poland's coal exports fall from 15 mtoe to 7 mtoe, with a further moderate decline as 2020 approaches.

The drop in coal exports is partly compensated on the national market by policy measures that secure a high share for coal in electricity generation, heat production, and industrial processes. Domestic coal consumption increases slightly from 2005, following a reduction in the decade after 1995.

In the other Visegrad countries (the Czech and the Slovak Republics and Hungary), hard-coal production continues to lose ground along with the decline in domestic demand. Residential and industrial use of hard coal drops in response to modernization and investment in new technologies. The poor reserve base for hard coal inevitably leads to a gradual reduction in underground coal mining. Costs of mining are for the most part considerably higher than the costs of alternative fuels, and they are rising due to unfavourable geological conditions. Employment considerations are taken into account in the process of shutting down mines, but they do not present a permanent hindrance to reforms. The major part of the reduction in employment took place prior to 1995.

In the United Kingdom production stabilizes at 25 mtoe on an annual basis, after having declined from 32 mtoe in 1995. The privatized British Coal

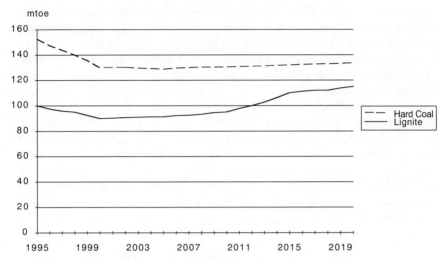

F<small>IG</small>. 10.8. Hard-coal and lignite production in Europe

manages to reduce coal-mining costs in line with the decline in international coal prices. The opportunities that exist in the United Kingdom for continued productivity improvements (primarily though a relaxation of working-hour restrictions and other labour benefits) are exploited. This is justified by reference to concern over increased dependence on energy imports.

Germany and Spain are the only European countries to maintain massive state support for hard-coal production. For both countries, production falls slightly in the late 1990s but stabilizes thereafter. Modernization and productivity improvements continue, but at a slower pace than in coal-exporting countries because of the difficult geological conditions. Thus, actual state support to the coal industry is rising.

Production of Lignite

The structural decline in the demand for lignite, particularly in the residential sectors, inevitably leads to a decline in production. However, this reduction is temporary, and European lignite production regains its 1995 level soon after 2010. In Central Europe, lignite remains an important source of energy in district heating plants, but power generation is the only application of lignite use that is growing. By then most of lignite production is used in pit-head power stations.

Lignite is the most important indigenous energy resource in Hungary, and in the Czech and Slovak Republics; it is also important in the energy balance of Germany. Lignite is therefore seen as an important means of curbing the growth in energy imports. Policies are installed to improve the productivity in lignite mining. Moreover, special efforts are made to reduce the environmental impacts of lignite mining and consumption. Despite this, environmental considerations restrain some of the growth in lignite supplies. Hence lignite is only a small portion of the energy consumption growth that accelerates in Central Europe after the turn of the century.

From 2010 lignite production starts to increase in the Czech and the Slovak Republics, Hungary, and Germany. This is a result of improved competitiveness *vis-à-vis* oil and natural gas, due to price increases for the latter, and progress in mining and combustion technologies for lignite. Rising import dependence encourages political support for lignite mining in countries with a large resource base.

District Heating and Combined Heat and Power

Apart from the power sector, district heating plants are the most important outlet for lignite. District heating retains its dominant market share in Central Europe, as governments continue to provide financial support to utilities and associated coal mines. In view of the poor financial performance of district heating companies, Western energy enterprises are not interested in entering into partnership agreements, nor are local district heating companies able to undertake large-scale rehabilitation. Energy efficiency in generation, transport, and end-use of heat remain low (see Box 10.3). The weak financial position of utilities and municipalities also hampers the introduction of alternative fuels, such as natural gas.

The continued public support to district heating is motivated by social and energy-security considerations. District heating is an important market for indigenous coal and thus helps to ensure employment whilst curbing imports of oil and natural gas—which are alternative fuels for energy-use in buildings. To raise heating tariffs to the level of supply costs is perceived as untenable and unfair since high supply costs are a legacy of the past regime.

In Western Europe, much less heat is supplied through district heating utilities, and there is only a moderate increase in its contribution to energy supplies to households, services, and industry. Pure district heating systems show no increase, whereas combined heat and power production (co-generation) rises quite rapidly in some instances. Countries with high capacity in co-generation, such as Denmark, Netherlands, and Italy, pursue a policy of continued political support for such schemes. In other countries where industrial co-generation

Box 10.3 *District heating in Central Europe*

A major part of energy needs for space heating and hot water in Central Europe is provided through district heating systems (heat-only boilers or combined heat and power plants). In urban areas more than half of all households are supplied with heat from district heating networks. Lignite and hard coal are the principal fuels for heat generation. Originally the motives for constructing large, centralized district heating systems were to eliminate local coal burning which entailed severe local air pollution. However the district heating plants are themselves major emitters of SO_2, NO_x, and particulates, causing both local and regional environmental damage. By the mid-1990s few plants were employing pollution-control technology, and energy losses in boilers and the transmission and distribution systems were generally very high.

District heating plants (and combined heat and power plants) are often located outside city centres. This has helped to reduce the effects of air pollution, but it has increased energy losses in transmission and distribution systems. Poor technological standards and inadequate maintenance have further contributed to the losses. Lack of heat controls and metering equipment have also contributed to major wastage in the end-use of energy in buildings.

During the Communist period, heating tariffs were extremely low. They rose in the early 1990s but still remain below costs. As a consequence, district heating companies have remained in a precarious financial situation, often unable to undertake badly needed rehabilitation.

based on natural gas growing rapidly at first (notably France, Germany, and Spain), growth declines as the economic terms for selling surplus power to the grid become less advantageous. This follows complaints and pressure from large utilities which claim that co-generators are selling nearly all of their power production to the grid at preferential rates.

The Power Sector

The Fuel Mix

Despite efforts in many countries to curb the growth in natural-gas imports, gas gains market shares in the power sector before 2010, primarily at the

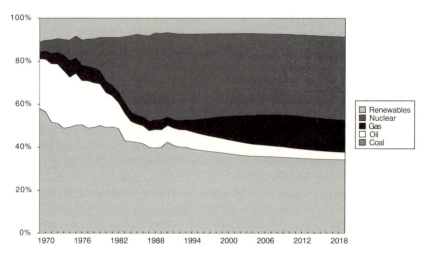

F<small>IG.</small> 10.9. Fuel shares in electricity generation in Europe

expense of coal. There is also a relative decline for nuclear and oil because few new plants are ordered between 1995 and 2005. However, the nuclear share starts to increase again after 2010. Still, the fuel mix changes only modestly compared to the radical shift during the 1970s and 1980s. The change—or rather the stability—from 1995 to 2020 is partly the result of active energy-policy initiatives and is partly brought about by technology developments and the pursuit of a reduction of costs in power generation and cut in emissions. Two aspects are of particular importance here: competition between natural gas and coal, and the future of nuclear power.

Natural Gas versus Coal

Natural gas became economically and environmentally attractive for power generation in the late 1980s. From 1990 to 1995 natural-gas use increased more than the total increase in primary-energy use, reflecting for a large part a contraction in coal use, particularly in Central Europe but also in the United Kingdom. The strong growth in natural-gas demand continues after 1995, at an annual rate of 10 per cent to 2005. There is stagnation in coal demand for power production despite an active policy to promote coal.

Germany has a sizeable increase in natural-gas demand despite its coal policy. From 1995 to 2005 most of the incremental fuel requirement is met by natural gas. This does not jeopardize the coal policy, since there is no target of increasing coal production. A similar development takes place in Spain, where electricity demand grows rapidly. Easier access to natural gas through

Europe—the Mahgreb Pipeline from Algeria—spurs investment in gas-fired power plants; meanwhile, plants based on indigenous coal continue to operate as before. Apart from being economically attractive, natural gas is preferred by South European utilities for environmental reasons. Both local concerns and policy commitments to cut SO_2 and NO_x emissions are important in this regard.

In Central Europe, fuel requirements for electricity generation continue to fall after 1995, due to sluggish growth in electricity demand and efficiency improvements in power supplies. Coal is particularly affected, due to its large market share and due to the major environmental problems related to the use of low-quality domestic coal. But, as explained above, domestic coal production continues to be supported and this hinders a steep decline in coal use for power production. The same is the case in Germany and Spain. Other South European countries and Ireland see an increase in coal demand due to rapid growth in electricity consumption, coupled with a policy to promote coal imports

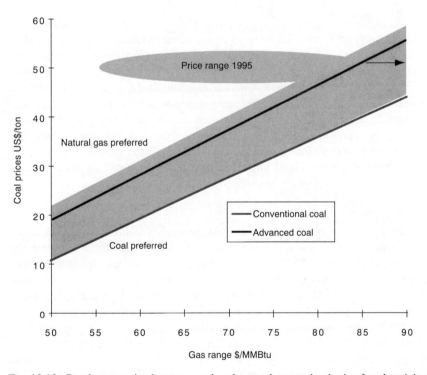

FIG. 10.10. Break-even price between coal and natural-gas technologies for electricity generation

because of energy-security considerations. Moreover, some areas lack infra-structure to support large growth in natural-gas supplies.

On the other hand, the economic attractiveness of combined cycle gas technologies over conventional coal technologies is overwhelming during the period 1995–2005; see Fig. 10.10. At a natural gas price of 2.7–3.0 $/MMBtu in 1995, and with only a modest increase before 2010, combined-cycle gas turbines enjoy a clear competitive advantage over coal. Prices for natural gas would need to be 30–40 per cent higher to make conventional coal technology the preferred option. However, it is not the current prices that are the appro-priate parameters for investment decisions, but rather price expectations over the lifetime of the investment. The gradual tightening of the natural-gas market raises price expectations, eventually leading to an upward trend in import prices for natural gas, see Box 10.2. Moreover, improved coal technologies come into commercial use after the turn of the century and greatly improve the commercial and environmental attractiveness of coal for power generation; see Box 10.4.

This explains the renewed growth in coal for power generation after 2005. From 2010 to 2020 some 60 per cent of the additional fossil-fuel requirements is coal. This shift is not only the result of the enhanced economic and environmental performance of coal technologies, it also follows as a result of greater concern over growth in natural-gas imports from Russia and Algeria. Since utilities in many countries are not exposed to competition they can 'afford' to take a long-term perspective on their investments and thus choose coal (or nuclear). In addition, governments and the EU, throughout the entire period from 1995 to 2020, allocate considerable funds to research, develop-ment, and demonstration of advanced coal technologies. This starts to bear fruit towards the end of the period.

Nuclear power

During the 1990s there is only a modest increase in nuclear generating capa-city, reflecting the standstill orders for new plants that followed after the Chernobyl disaster in 1986. However, national energy-security concerns spur a new round of nuclear plant construction. Orders for new plants increase gradually after the turn of the century, parallel to the increase in European natural gas imports. Due to the relatively long lead-time from plant order to available production, it is not until 2010 that nuclear production increases significantly. But during the decade from 2010 to 2020 European nuclear production increases from 925 TWh/year to 1050 TWh/year. This increase represents an additional capacity of close to 20 GW. The major part of the capacity additions come in countries which at the outset have a large nuclear

Box 10.4 *Coal combustion technologies*

Thermal efficiency and environmental technologies have improved considerably in recent years. Further improvements are foreseen in the next few decades; cost reductions from the technologies and for coal will improve the commercial attractiveness of coal for power generation, even with stringent environmental standards for dust, SO_2, and NO_x. Three categories of technologies are of particular importance for the next two or three decades.

Pulverized Coal (PC) based on conventional steam turbine technologies may soon after the turn of the century operate with a thermal efficiency of close to 50 per cent. Plants in Japan have from the early 1990s attained an efficiency of more than 40 per cent. Pulverized coal with appropriate control technologies may remove considerable parts of SO_2 and NO_x emissions.

Fluidized Bed Combustion (FBC) covers two advanced coal combustion technologies: Atmospheric Fluidized Bed Combustion (AFBC) and Pressurized Fluidized Bed Combustion (PFBC). AFBC is commercial in the mid-1990s, though with a thermal efficiency generally below 40 per cent. PFBC, which will not be commercial until after 2000, has the potential of higher thermal efficiency (exceeding 50 per cent) and the advantage of smaller size and thus lower capital costs. Both AFBC and PFBC can remove SO_2 up to 95 per cent and NO_x up to 80 per cent. Both FBC technologies, especially PFBC, have great fuel flexibility, so that low-quality indigenous coal can be used.

Integrated Gasification Combined Cycle (IGCC) may attain a thermal efficiency higher than PFBC and remove practically all SO_2 and up to 90 per cent of NO_x. IGCC exists in the 1990s only in demonstration plants and appears to be a longer term option than FBC. A coal gasifier feeding a fuel cell (conversion of energy to electricity through a flawless oxidation reaction, as in a battery) is even more distant as a commercial option but it does signify that coal is not an energy source of the past.

The future capital costs of the PFBC and IGCC, not yet in commercial use, are uncertain. Optimistic estimates indicate costs of about $1,100 per kilowatt in 2010, compared to $800 per kilowatt for combined-cycle gas turbines in the mid-1990s (all in constant dollars).

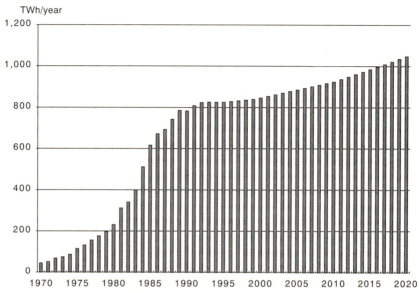

Fɪɢ. 10.11. Electricity from nuclear plants in Europe

share in their electricity balance; i.e. France, Germany, and Spain. Even with the expansion of nuclear production in these countries their reliance on natural gas from Russia and Algeria increases markedly.

National energy authorities and the European Union are not only active in promoting the construction of new nuclear capacity, major efforts are also made to maintain a high utilization rate on existing and ageing nuclear plants. After 2000 many nuclear plants are approaching their stipulated economic lifetime and large investments are needed to keep them in operation with satisfactory production regularity and safety standards. The decommissioning of plants is kept at a low level throughout the entire period from 1995 to 2020. It is becoming economically attractive to refurbish old plant. Total refurbishment of nuclear plant started with Sweden's Oskarshamn plant, which from 1992 to 1995 was furnished with a new reactor (entirely replacing an old reactor, which had never been done before) at an investment cost recovered in ten years.

11
Liberalization and Trade

This scenario depicts a future with extensive political co-operation and integration in the EU, and where the Visegrad countries become EU members. Restrictions on trade and movement of capital and labour in Europe are gradually abolished, first in Western and subsequently in Central Europe. Industries previously shielded from competition have to restructure or close down. Trade liberalization and an economic environment conducive to foreign investments are pursued to foster economic growth and job creation. The policy of competition and open trade is also extended to the energy sector. In Central Europe, the commercialization and privatization of energy entities gain momentum. In the West investments in grid-based energy sectors are increasingly made on commercial criteria, and market pressures contribute to the dismantling of vertically integrated supply monopolies. Major parts of the energy sector become subject to competition, and energy costs and prices converge between countries, even for electricity.

These changes have major implications for energy-consumption trends. Energy-sector reforms represent radical changes to the framework within which the energy industry operates. The shift from state intervention and control to market orientation, and from national to supra-national policy formulation and implementation, has far-reaching consequences for the electricity sector. The liberalization of coal and natural-gas markets gives further impetus to this development. Therefore, it is the electricity sector that is the main focus of this chapter.

11.1. Main Features

Revitalization of European Integration

After the temporary set-back in the integration efforts that followed the signing of the Maastricht Treaty, the latter part of the 1990s initiates a sustained trend of closer political and economic integration in Europe. Important steps are taken to reform governance and institutions within the EU. In short, this implies that member-states agree to cede national sovereignty over a number of policy issues. Membership negotiations are initiated with the Visegrad countries, and by 2000 they have become full members. Initially, a selected list of 'sensitive' products is excluded from full competition, and a few sectors in the Visegrad countries are granted temporary exemptions from EU regulations. However, soon they have to comply with the competition rules of the Rome Treaty and other EU regulations.

The Internal Energy Market

The EU Commission finally gains political support for the decade-long struggle for increased competition and trade in energy. This is made possible not only through the increased power of EU institutions, but very much as a result of initiatives, launched at national level in several member countries, to deregulate and increase competition. The EU reforms focus primarily on the grid-based energy sectors 8 natural gas and electricity. These grids are opened for third-party use—at first only to a limited number of large customers, but as regulatory systems are developed, and national energy enterprises adapt to the new market conditions, competition is broadened.

Trade liberalization in electricity and natural gas calls for all-European regulations and regulatory bodies to oversee the operations of natural monopolies and other entities in the energy market that potentially may abuse a dominant market position. The EU Commission is given the role as supervisor in this regard.

EU policy, supported by the general rules of the Rome Treaty, also has a major bearing on public support to energy industries, primarily state aid to coal production. Coal subsidies are dismantled across member countries, with radical consequences for indigenous coal production. Governmental support and subsidies to existing nuclear plants are reduced, and support to new nuclear projects discontinued.

A Common Energy Policy

The basic political rationale for a common EU energy policy is to install and supervise a regulatory system that ensures economic efficiency in energy supplies and to address the concern for security of supplies. With the dwindling supplies of domestic energy in many European countries, including the closure of coal mines and the gradual phase-out of nuclear, there is greater dependence on energy imports from politically unstable areas. Securing reliable energy supplies therefore becomes an issue of growing political and public concern.

Energy policy is lifted from the national level to the EU level, but direct policy interventions are restricted by the commitment to maintain competition and a 'level playing field' for the energy industry. However, EU policy has some effect on fuel choices, particularly within the power sector. Large funds are allocated to restructure hard-coal and lignite production in Central Europe. Support is also granted to ensure safety and operational efficiency of nuclear plants both in Western and Central Europe. This postpones the decommissioning of nuclear stations approaching the end of their economic lives. Various measures are also implemented to encourage investments in renewable energy sources.

Regulatory Framework

The new energy-sector paradigm sets out a system of clearly defined roles for commercial companies, regulatory bodies, and policy institutions. In the old regime the demarcation lines between commercial, regulatory, and policy functions were blurred or often non-existent. Governments accomplished control through direct and indirect regulatory measures on vertically integrated and/or government-owned or -controlled energy industries. The new structure requires an entirely new regulatory framework. This is often (and somewhat misleadingly) labelled 'deregulation'. Policy, regulatory functions, and commercial operations are clearly separated. In some countries extensive legal and administrative systems are developed, and in all countries, the role of regulatory bodies is strengthened.

The reason for this 'upgrading' of regulatory functions is a transformation to a system with more actors in the market that operate independently of each other and often in direct competition. Whereas the previous structure was largely self-regulating, with one or a few entities ensuring coordination and stability in exchange for exclusive rights in the market, the new system calls for sophisticated arrangements in order to guarantee efficient coordination of all stages of energy supplies, and a transparent and active regulatory system to ensure fairness and efficiency.

Energy-industry Structure

The new market conditions emerging from the regulatory changes have major implications for energy-industry structure. In many countries the new competitive business environment leads to concentration in larger units with the financial strength and technological ability to compete in the international market-place. In Central Europe, energy-sector entities are established as joint-stock companies and often privatized. Enterprises all over Europe

become increasingly international in their strategies and operations. This helps
to spur private-sector investment in the energy sector in Central Europe.

The new commercial and international orientation of the energy industry
triggers a series of mergers, joint ventures, and cross-ownership among energy
enterprises. Some companies shift from being national companies to becoming
all-European in their operation, with considerable involvement elsewhere as
well. The trend towards concentration is partly counterbalanced by regulatory
changes. First, the natural monopolies in energy transmission and distribution
are subjected to strict regulation which requires that such activities be established as separate commercial and operational entities. Secondly, the dismantling of the vertical structure of energy enterprises helps to facilitate the entry
of new, independent, and often small energy suppliers to the sector. This in
turn is greatly helped by new technologies that make small-scale operations
commercially attractive.

New Technologies

Technological change plays an important role in energy developments to 2020,
despite the inertia of such a capital-intensive sector as energy. The radical
regulatory changes described above and the structural changes accompanying
them are made possible through technological innovation. Without the new
information technology, a radical deregulation and decentralization in the grid-
based energy supplies would not have taken place. The growing market share
for independent power production and co-generation of heat and power is the
result of new gas combustion technologies adopted from aerospace technologies. Economic integration also accelerates the scrapping of old capital equipment and promotes the dissemination and introduction of efficient and
environmentally benign new technologies. Moreover, electricity-sector
reforms facilitate the entry of independent power producers and improve the
competitive position of small-scale production, including new renewable
energy sources.

Technologies Promoting Gas Use

The new developments in regulatory regimes and technologies mean that
small-scale energy-supply systems in the power sector will often have a
competitive edge over larger and more capital-intensive options. Low capital
costs, short construction time, transparent and relatively lenient licensing
procedures, and environmental advantages work in favour of gas. However,
it is important to take into account the fact that the attractiveness of different

technologies in power generation will often hinge on expectations of future fuel prices. Comparing investments in coal- or gas-fired capacity, for example, the economic attractiveness of the options is highly sensitive to expected price trends. Rising gas prices and the continued decline in coal prices encourage R&D and an eventual commercial breakthrough for advanced coal combustion technologies.

11.2. Energy Supply Industries: Reforms and Policies

The Sequence of Sector Reforms

With the oil markets of most European countries liberalized, three sectors are from the mid-1990s onwards the subject of radical reforms: coal, natural gas, and electricity. Technically and politically, coal is relatively uncomplicated to liberalize, employment considerations being the major obstacle. Reforming the natural-gas sector entails greater technical challenges and more sensitive political issues. Reform proposals are met with considerable resistance from major actors in the gas industry and national energy authorities, and the merits of radical reforms, in terms of economic efficiency, are questioned. The same is the case for electricity; here the implications of market reforms are potentially more pervasive, due to the dominant role of electricity in the energy-supply systems of Europe.

The scope and timing of reforms of the three sectors differ. By the turn of the century the remaining coal mining in Europe is operating without financial transfers or other market protection. Changes in natural-gas markets come more gradually: first with a 'negotiated third-party access' in the latter part of the 1990s and subsequently with a fully fledged third-party access from 2005. In electricity the process is even more cautious, due to the technical coordination and commercial viability needed of national electricity enterprises. An integrated and liberalized electricity market in Europe is not fully in place before 2010.

There follows a discussion of the main features of this reform process, and the new supply patterns and regulatory systems that emerge.

Natural-gas Sector Reforms

The New Industry Structure

Reforms impact primarily on the downstream part of the market, first of all through an alteration in the regulatory and institutional framework for

transmission and distribution of natural gas to end-users. The gradual introduction of third-party access facilitates the emergence of new players in the
gas markets, and the role of the existing importers and owners of the transmission grids undergoes major changes. In countries with a system of independent
distribution companies (e.g. Italy, Spain, the Netherlands, Belgium), the introduction of third-party access gives rise to new alliances between gas distribution companies. They see an interest in developing gas-trading bodies that
challenge the position of established importers. Genuine trading houses for gas
are established, partly as independent enterprises, by industrial companies or
electricity producers and as joint ventures between distributors and industrial
gas-users and electricity producers.

Power companies and manufacturing industry are the first to reap the
benefits of liberalization. For smaller users the immediate benefits are more
modest. However, as market liberalization proceeds, with a continued market
and political pressure against price discrimination and the dominance of
vertically integrated companies, there is a tendency towards convergence in
gas prices for similar supply contracts. Although there is a clear trend towards
harmonization of the structure of the gas industry across Europe, the established gas companies react to the competitive pressures and adapt their pricing
and sales policies in order to maintain a hold on their markets. Instead of being
protected by legislation they attempt to protect their markets through a more
flexible market behaviour.

A further change in the institutional structure results from producers becoming more active in downstream activities in the gas market. They establish their
own marketing companies to sell gas to large users in industry, to electricity
generation, and to distribution companies. This phenomenon was already
apparent in the first part of the 1990s in the United Kingdom, following the
dismantling of the monopoly of British Gas and the opening of the UK grid to
third parties.

Price Effects

Third-party access has implications for price levels and structures throughout
Europe. First, cross-subsidization between the various categories of users and
existing price differentials for large gas users are gradually eliminated. Some
of these differentials were due to differences in efficiencies of the gas utilities,
and some were the result of the application of the 'market value' principle.

Downstream competition in the gas markets results in a squeeze on the
earnings of some of the established gas companies. Gas-to-gas competition
gives new players on the market the possibility of 'cherry picking', i.e. con-
centrating their marketing efforts on the most profitable parts of the market.

'Cherry picking' eliminates higher than normal earnings in individual segments of the market. Gas companies are therefore under pressure from the
market to adapt their pricing policies and gradually adjust price levels to reflect
the real costs of supplying the various groups of gas users. The result is lower
prices for large industrial customers and power companies.

Third-party access also spurs competition between producers of gas. However, the gradual pace of the sector reforms and the small number of suppliers
capable of bringing new supplies to the market means that gas-to-gas competition emerges slowly and with only modest downward pressure on import
prices. Other forces also work against lower gas prices. Third-party access
increases to some extent the uncertainties in the gas market. With new market
players gaining access to the market, the privileged position of the major
importing and transmission companies is eroded. They lose market power
and are thus less willing to undertake long-term contract commitments. This
penalizes the most capital-intensive gas-supply projects and eliminates the
development of some supplies.

In summary, liberalization gives a downward pressure on prices in the short
and medium term as a result of gas-to-gas competition and rationalization (and
occasionally lower profit margins) in transmission and distribution. Beyond the
year 2005, however, prices show an upward trend as requirements for new
sources of supply grow.

Dismantling Coal Subsidies

As international coal prices continue to decline, European coal production
becomes increasingly uneconomical and thus demands reform in order to
improve public finances and industrial competitiveness. Employment considerations alone do not justify continued and increasingly costly support to
domestic coal production, and energy security is no longer an argument in
favour of protecting the coal industry. As a result there is a major contraction
in European coal production throughout the 1990s and during the first years
after the turn of the century.

Commercialization and Privatization

In Central Europe the process of enhancing the business orientation of the
coal industry started in the early 1990s and proceeded at a rapid pace from
the mid-1990s. Enterprises are given autonomy in matters like pricing, bor-
rowing, staffing, and procurement, and subsequently made subject to company
legislation, which is a precondition for privatization. By the turn of the
century, all coal-sector activities in Central Europe are in private hands. In

Poland, hard-coal production is transformed from a centralized industry structure, of the type prevalent under the Communist regime, to a structure of several independent companies, in order to avoid monopolistic conditions in the domestic market. Similar structures are established in the other Visegrad countries, but the liberalization of energy markets implies that hard-coal production in these countries is drastically reduced. In the United Kingdom all hard-coal production is undertaken by privately owned British Coal, which competes with coal imports, fuel oil, and natural gas, mainly in the market for power-sector fuels.

District heating systems, most common in Central Europe, are primarily kept as public utilities. They are allowed to operate without policy interference but are kept under constant observation by regulatory authorities in order to avoid abuse of a dominant market position. District heating companies have to compete with other energy suppliers for market shares and do not benefit from any politically installed price protection. Nor do they enjoy financial transfers from state budgets. This implies that the district heating sector undergoes major changes which result in rationalization and a decline in the total scale of large district heating schemes.

Hard-coal Production

Dismantling of subsidies to the coal sector has radical implications for hard-coal production. Only in Poland and the United Kingdom is production viable without governmental support. As a result, total European hard-coal production is reduced from 156 mtoe in 1995 to 84 mtoe in 2005. Production in 1970 was 263 mtoe. In the ten years from 1995 to 2005 employment falls by 60 per cent to 200,000. Although this is a major drop in relative terms, the job losses in the preceding ten-year period were higher. In 1985 some 800,000 were employed in European coal mining, against 450,000 in 1995. At the same time as financial transfers to the coal sector are discontinued, restrictions on coal imports are lifted, abolishing the purchase obligations for domestic coal by utilities and steel producers. Energy consumers are thus free to import less expensive coal or shift to other energy sources such as natural gas or fuel oil.

The most radical reduction in coal production takes place in Germany. The last mines are closed before 2005. The best performing German coal mines have production costs considerably above the cost of coal imports. Spain follows a similar path, with most production being closed soon after the turn of the century. In France production is discontinued by 2005 in line with a governmental decision from 1993. In the Czech and the Slovak Republics and in Hungary most hard-coal production is closed before 2005, for economic reasons. After 2005 the only remaining hard-coal producing countries are

Table 11.1. *Hard-coal production in Europe, 1970–2020 (mtoe)*

	1970	1995	2000	2005	2020
Poland	78	62	60	60	65
Germany	73	43	15	0	0
United Kingdom	75	30	20	18	16
Spain	7	7	2	0	0
Other countries	30	12	8	4	4
Total Europe	263	154	105	84	87

Poland and the United Kingdom. Liberalization of the British energy sector had the greatest impact on coal production before 1995. Privatized British Coal manage to contract deliveries to the electricity-supply industry of 25 tonnes/year from 1997 and further deliveries of 3–5 tonnes/year to other domestic sectors. However, as international coal prices continue to decline it becomes increasingly difficult for British Coal to stabilize supply volumes in competition with imported coal and heavy fuel oil. Hence there is a slide in hard-coal production from 32 mtoe in 1995 to 16 mtoe by 2020.

The Polish coal industry saw the most radical contraction in production prior to 1995. The economic recession from 1988 to 1994 resulted in a 35 per cent drop in coal demand. The economic restructuring of Poland's coal industry continues at a rapid pace after 1995 with considerable improvements in productivity. Employment is reduced from 300,000 in 1995 to 125,000 by the turn of the century, after having been 430,000 in 1988. This restructuring makes much of the Polish coal industry competitive with energy imports, and domestic coal remains the principal energy source in the Polish energy balance. Coal exports, on the other hand, continue to fall. The differences in import and export parities that give geographical protection for domestic coal also mean a low netback profit for coal sold on the international market.

Lignite Production

Lignite mining is, for a large part, economically viable without state protection. Hence the decline in production is fairly modest from 1995 to 2005, and there is a rebound in lignite mining after 2010, as higher prices for oil and natural gas make production more profitable.

Germany has the largest reserves and best production potential in Europe. Production in the old Länder remains competitive at about 22 mtoe, and with somewhat higher output after 2010. In the new Länder, where production dropped from 63 mtoe in 1989 to 20 mtoe in 1995, production stabilizes at about 15 mtoe from 2000 to 2010, with a subsequent 20 per cent increase

Table 11.2. *Lignite production in Europe (mtoe)*

	1970	1995	2000	2005	2020
Germany	78	42	35	35	43
Poland	9	14	12	10	10
Czech and Slovak Republics	18	20	10	10	12
Hungary	6	6	2	1	2
Other countries	5	16	16	16	20
Total Europe	115	98	75	72	87

towards the year 2020. In other Western European countries, notably Greece and Spain, production remains more or less stable at 16 mtoe until 2010. Production above this level is hindered by environmental considerations, but after 2010 there is a slow increase as other fuel prices and import dependence grow.

In Central Europe about half of lignite production in 1995 is for non-power-sector use, for district heating plants and industry, whereas practically all use in Western Europe is for electricity generation. Restructuring of industry and the district heating sector results in a sharp decline in lignite use for purposes other than power generation. Environmental pressures work in the same direction. The privatization of lignite power plants reduces their share in the electricity market, notably in Hungary and the Czech and Slovak Republics. More stringent emission standards and environmental problems at or near the mining sites, including air pollution and waste disposal, also reduce the economic attractiveness of lignite as an energy source for power generation. Still, many of the lignite-based power plants that were in production in the early 1990s are refurbished and continue to operate within a competitive power-generation environment.

Electricity-sector Reforms

Due to the technical complexity and pervasive nature of electricity supplies, liberalization of this sector proceeds at a gradual and cautious pace. However, over a ten to fifteen year period from 1995 there is a radical transformation in the corporate structure and business orientation of the electricity-supply industry. Electricity is regarded as a commodity to be offered in a competitive market, and enterprises are concerned with short-term profitability and cost reductions as well as with long-term strategic investments in the international market.

The difficulties in establishing an integrated electricity market in continental

Europe, with the necessary ability to ensure economic efficiency and technical reliability, are given due consideration. Political factors are also important for the gradual approach. Much of the initiative for reforming the sector lies with the national authorities, whose approach to the issue depends on the initial structure of electricity industry and the associated fuel markets. The speed at which national electricity markets are opened for competition in electricity generation and in the retail market also varies. For governments it is important to proceed in a way which will enable the major domestic enterprises to consolidate their financial position and be prepared for the new competitive market regime.

Privatization and the free flow of capital give the most powerful electricity enterprises the opportunity to grow even stronger. The major German and British companies, and EdF of France, take large shares in the electricity industries in Central Europe. Shares in electricity enterprises in Western European countries are also open for purchase. At the same time there is a marked trend towards ownership concentration in all European countries which have had a decentralized structure. This is supported by national energy and competition authorities, in order to prevent a few major enterprises from attaining a dominant position in the electricity market. Competition authorities closely follow the development of ownership concentration and take steps to hinder mergers in cases where competition is in jeopardy.

In Chapter 6 we described the considerable variety in the regulatory systems and institutional structures of the national electricity markets in Europe and how reform processes from 1989 to 1995 actually moved in different directions. The liberalization of the electricity markets across Europe that starts after 1995 reverses this trend, and gradually leads to a convergence in structures between countries. This takes place over a period of more than ten years, and not until 2010 can this process be considered finalized. With the reforms in Europe going in the same direction, there are still prevailing differences in electricity structures.

The new structure of the electricity sector has a major impact on energy markets. Generally, costs are reduced as the 'cost-plus regime' is no longer applicable under the new market conditions. There are, however, other factors that work in the direction of higher prices. Tariff structures are altered as cross-subsidization practices are discontinued. Fuel choices are affected by operational dispositions which in the liberalized market often differ from earlier regionalized markets. Investment and disinvestment decisions based on commercial criteria, together with the liberalization of the fuel markets, have major impacts on the demand for primary-energy commodities such as coal, oil, and natural gas. This is examined further at the end of the next section.

11.3. Energy Consumption Trends

Demand Determinants

Integration and trade affect the economy in three ways that are important for energy demand trends.

1. *Economic growth.* Economic integration and trade give benefits in the form of higher economic growth rates. In Western Europe the benefits are temporary and exhausted by 2005, whereas in Central Europe there is a high economic growth rate throughout the period to 2020; see Box 11.1.

2. *Changes in the composition of economic activity.* Trade liberalization, free movement of capital, the elimination of government influence in industrial management (e.g. for social or industrial policy reasons), and lifestyle changes lead to major shifts in the composition of economic activities. These changes have a significant influence on energy demand.

3. *Technological progress.* Liberalization and the elimination of trade barriers foster the development, introduction, and market penetration of modern

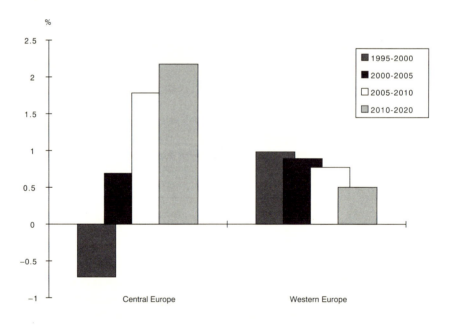

Fig. 11.1 Total energy consumption in Western and Central Europe, 1995–2020 (average annual growth rates)

Box 11.1 *Benefits of ' European Integration'*

After the Single Act was ratified in 1987, many studies have attempted to estimate the welfare effects of implementing the 'four freedoms'—free movement of goods, services, labour, and capital. Studies have focused on reduced costs/margins due to dismantling of trade barriers, due to economies of scale, and due to stronger competition, and the welfare effect due to non-price factors. The EC Commission's own study indicated economic gains ranging between 3 per cent and 7 per cent of GDP. Estimates from other independent studies for the most part fall within the same range (see Cecchini, 1988; Hoeller and Louppe, 1994). The benefits of integration and trade are most likely to be greater and more durable for the countries of Central Europe, because of large inefficiencies in most economic sectors.

In the liberalization and trade scenario a GDP growth rate in Central Europe between 1995 and 2005 is asssumed to be 0.5 percentage points higher than in national rebound. From 2005 to 2020 this gap narrows to 0.15 percentage points. By 2020, the GDP level of the liberalization and trade scenario is 8 per cent above the level in national rebound. In Western Europe, the benefits of liberalization are of a more temporary nature; the difference in the GDP level in 2005, when benefits in terms of higher growth rates are exhausted, is 2.5 per cent.

Average annual growth rates of Gross Domestic Product

	1995–2005		2005–2020	
	Liberalization and trade	National rebound	Liberalization and trade	National rebound
Western Europe	2.8	2.6	2.6	2.6
Central Europe	5.0	4.5	4.5	4.3

technology. Enhanced productivity in industry promotes investment in new production processes and accelerates capital-stock turnover. The industry's search for productivity gains and competition on open markets fosters the development of new technologies to reduce input costs, including energy, and enhance product quality. Consequently, technological progress leads to more energy-efficient equipment and production processes.

These three factors impact on energy demand in both Central and Western

Europe, with a continuous downward trend in energy consumption growth in Western Europe and a clearer upward trend in Central Europe.

Sectoral Energy Demand

Industry

The unrestricted flow of capital and labour together with increased competition leads smaller industries to merge and thereby establish larger and more power-ful companies, better suited to compete on open markets. In addition to pressure from competition, mobility in capital flows helps to close the tech-nological gap between industries in different countries, in particular between East and West.

The industry structure also changes because of relocation to regions outside Europe. Stringent environmental regulation and rising costs for labour and energy affect energy-intensive industries in Western Europe. These industries improve their competitive position through migration to low-cost regions. Though liberalization and trade foster industrial migration, a significant indus-trial base still remains in Europe, without which economic growth would not be possible.

Despite a marked contraction of energy-intensive industries in Central Europe, industrial output generally maintains a significant share in industrial output. This is caused by a high level of private and public investment, which keeps up the demand for machinery, equipment, and construction material. Moreover, as trade restrictions are eliminated, certain energy-intensive indus-tries in Central Europe (such as steel) increase their sales on the integrated European market. There is also an increase in the car and electronic industries of Central Europe. All these industries enjoy competitive advantages because of lower costs for some inputs, such as labour, and less stringent environmental legislation, well into the twenty-first century. In addition, prospering markets spur the migration of industries to these countries. The growth in industrial production is therefore considerably higher in Central Europe compared to Western Europe (see Fig. 11.2), despite the relatively small difference in energy demand. This, in turn, implies that there is a sharper fall in industrial energy intensity in Central Europe. In 2020 energy consumption per unit of value added in Central European industries is half the 1995 level; in Western Europe the reduction is 35 per cent.

Similar to the developments of the 1980s described in Chapter 2, the reduc-tion in industrial energy intensity can be separated into two elements: struc-tural change and energy-efficiency improvements (see Fig. 11.3). Both factors

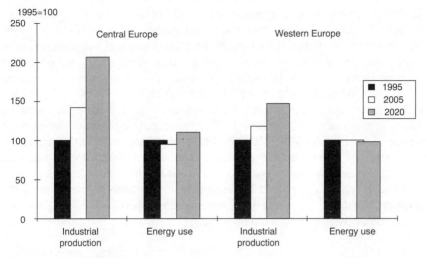

Fɪɢ. 11.2. Industrial production and industrial energy consumption in Central and Western Europe (1995 = 100)

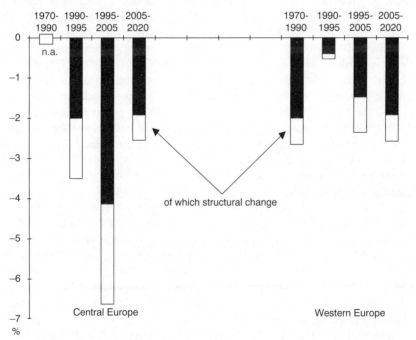

Fɪɢ. 11.3. Industrial energy intensity in Central and Western Europe, 1970–2020 (average annual growth rates)

contribute to the reduction in and stagnation of industrial energy demand. Structural changes are most important from 1995 to 2005. In Western Europe this is driven by higher energy prices for some industries and elimination of state protection, leading to the relocation and close-down of 'old industries'. In Central Europe changes in industry structure are a continuation of a trend which started before 1990 and peaked between 1990 and 1994.

Energy efficiency in Central Europe improves through the replacement of obsolete machinery and equipment. This effect was also present earlier but it grows stronger after 1995 as industrial investment accelerates. There is also a continuous improvement in energy efficiency in Western Europe, in line with general technological progress and determined by the pace of replacement of old machinery and equipment. This increases requirements for electricity at the expense of labour, but at the same time reduces direct use of fossil fuels as production processes become more efficient.

Transport

European integration means increased freight-transport requirements both within the Western part of the European Union and due to the incorporation of Central Europe into the European trading system. As a consequence of the EU policy to expand the integrated European networks and to improve East–West links, there is a major increase of trade flows between Central and Western Europe. East–West trade gains an importance similar to that of the North–South route in Western Europe in the 1980s. Trade volumes also increase as a result of industry's efforts to increase productivity (reduction of storage capacity and exploitation of variations in certain cost factors, such as labour, within an integrated Europe). The pressure on additional transport capacity in the East is met by the construction of new roads. This leads to a decline in the freight volumes shipped by rail, which by 2010 approach Western European levels.

In Western Europe, rail transport increases its market share, mainly as a result of public investment in large-scale infrastructure which enhances the rail system. These efforts, driven primarily by environmental concerns, help to alleviate bottlenecks in North–South transport. This increases the share of electricity used for transport purposes, but the rail transport share of total energy use remains below 4 per cent for Europe as a whole.

Despite increasing freight-transport requirements and continuous growth in personal disposable income in Western Europe, there is a clear deceleration in the growth rate of transport fuel compared to the period from 1970 to 1995. This is, as explained in Chapter 10, the result of saturation effects. Still, the

1995=100

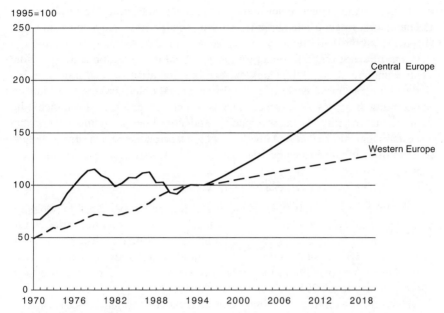

Fig. 11.4. Oil consumption for transport in Central and Western Europe, 1970–2020 (1995 = 100)

transport-sector share of oil consumption increases from 45 per cent in 1995 to 50 per cent by 2020.

Central Europe experiences a steady increase in transport fuel consumption of 3 per cent per annum throughout the period. By 2020 the level of consumption is more than twice that in 1995. Western Europe saw a similar increase from 1970 to 1995; see Fig. 11.4. Although the transport sector is the driving force for oil consumption in Central Europe, its share of total oil use remains below 50 per cent in 2020. The reason is increasing oil use also for heating purposes in industry and other sectors.

Service Sectors

Energy demand in the service sector is predominantly influenced by two factors: energy demand for space conditioning (heating and cooling) and electricity consumption in electric appliances (mainly office equipment and lighting). Both Western and Central Europe see a rapid increase in the demand for electricity, due to the spread of modern office equipment and uninterrupted increases in the economic activity of service sectors. Though saturation is reached in some office appliances, this does not lead to a stagnation in electricity demand because new appliances are introduced. In Western Europe, the demand for new office space is offset by productivity gains and new work practices.

Central Europe embarks on a process of radical reorientation and expansion of its commercial and public-service sectors. Liberalization and trade require that Central European economies be able to provide the services essential for undertaking business in an open society (banking, retailing, insurance). In addition, increasing disposable income and consumer spending necessitate the establishment of an acceptable level of commercial services, such as hotels and facilities for leisure activities.

Residential Sector and District Heating Systems

The positive impact on disposable income of trade and liberalization points to a rapid growth in the purchase of electrical appliances in households in Central Europe. This affects electricity consumption in particular. But the overall growth in energy consumption in the residential sector is lower than in service sectors and transport, due to the large improvements in energy efficiency for space heating from 1995 to 2005.

Rising end-user prices and improved financial performance of the energy industry allow energy utilities in Central European countries to refurbish their inefficient district heating systems. Through cooperation with Western partners, utilities gain access to modern combustion and heat transmission technologies, thus realizing substantial efficiency gains. In cases where the existing state of district heating systems is too poor to make refurbishment economically feasible, which is the case for about a fourth of supplies in 1995, natural gas gains market shares. District heating companies which remain unprofitable even under a liberalized price regime cease operation. Growth in disposable income leads to a substitution of coal by light fuel oil where the introduction of natural gas is not economically feasible, primarily in rural areas.

In Western Europe, energy use for space heating and cooking shows only a moderate growth whereas use of electricity increases significantly. There are, however, some signs of saturation towards the end of the period.

Electricity Sector: Electricity Consumption and Fuel Requirements

Consumption Growth

Electricity consumption grows more than other components of energy use, as it gains market shares in all major end-use sectors. Industrial processes increasingly require electricity for mechanical work and for process heat. Industrial restructuring leads to a higher share of energy requirements being met by electricity. On the other hand, there is a contraction in energy-intensive industries that helps to dampen the growth in demand. As described above, there is a persistent shift towards electricity in the residential and service sectors. Despite these developments, electricity growth rates are on a downward trend, from 3 per cent per annum from 1970 to 1995 to 1.3 per cent from 1995 to 2020. This deceleration in the growth of electricity consumption in Western Europe occurs despite the lack of a major Europe-wide increase in electricity prices.

Costs and Prices

Several conflicting factors influence electricity prices, giving higher prices in some countries and lower prices in others. First, the supply–demand balance in the electricity market becomes increasingly tight as the surplus generating capacity of the 1980s and 1990s disappears, due to low investment in new generating capacity. Commercialization of the electricity industry implies that investments in new generating capacity will not be forthcoming before electricity prices appear to yield a satisfactory rate of return. Market reforms mean that electricity prices are determined by the costs of bringing additional electricity supplies to the market. For many countries this is a break with earlier price-setting principles and it initially gives an upward shift in price levels.

Liberalization of the electricity market contributes to a significant downward pressure on prices in some countries. The elimination of coal subsidies in Germany and Spain has notable effects on generating costs in the two countries (see discussion below). Competitive pressures in power generation and distribution also yield sizeable productivity improvements which are largely passed on to electricity consumers. Moreover, capital costs of new supply capacity are on a downward trend due to technological improvements and increased competition among electrical equipment manufacturers. Streamlining of licensing procedures for capacity expansion and a general trend towards less capital-intensive solutions, implicitly with fewer controversies over siting of plants, further contribute to a slow but stable reduction in supply costs. Finally, increased trade in electricity improves the utilization rate for existing generating capacity and thus lowers supply costs.

The factors that contribute to lower prices are generally stronger after 2005 than before, since it takes some time for the full benefits of electricity reforms to gather speed. After 2010, however, there are only modest cost reductions from market liberalization, and higher fuel costs give a slow increase in generating costs and consumer prices. By the mid-1990s price levels vary substantially between countries and sectors (industrial and residential tariffs);

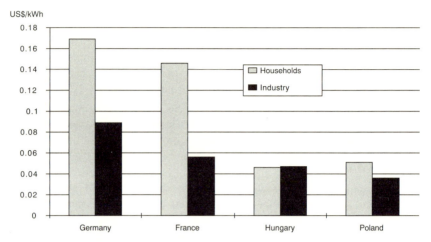

F<small>IG</small>. 11.5. Electricity prices in selected countries by sector, 1994

see Fig. 11.5. Both household and industrial tariffs are higher in Germany (including 'Kohlepfennig' which is abolished from 1996, concession fees, and stringent environmental standards) than in France, due to higher generating costs and earlier idle generating capacity which has landed French industry with favourable contracts for electricity supply. In Central Europe prices recorded in 1995 were generally low and prices for households did not cover the costs of supply. These price levels are gradually harmonized. In Central Europe, electricity price levels for households increase substantially even before the turn of the century; other adjustments are slower, and generally evolve as restrictions on electricity trade are relaxed and eventually eliminated.

Fuel Requirements

Fuel requirements of the electricity sector show less growth than electricity production and consumption, due to a continuous reduction in energy conversion losses. Modern coal- and gas-fired power plants are constructed with a thermal efficiency considerably above that of existing plants. Modern coal-fired plants have in 1995 an efficiency of about 45 per cent, and advanced coal-fired technologies which come into commercial use after the turn of the century attain an efficiency of more than 50 per cent—see Box 10.4. This compares with an efficiency of about 35 per cent for coal-fired plants made operational around 1980. Gas-fired combined-cycle plants, commercial already from the early 1990s, can obtain an efficiency above 50 per cent. Technology yields further improvements into the twenty-first century.

For this reason, the total increase in European primary-energy use in the electricity sector increases by 25 per cent from 1995 to 2020, against an increase in electricity output of 38 per cent. This means that the average efficiency in electricity generation rises from 37 to 41 per cent. But with the improvement in efficiency, electricity still represents the main factor of additional energy use from 1995 to 2020. Some 40 per cent of the growth in primary-energy consumption from 1995 to 2020 is attributed to the power sector. Moreover, fuel choices of the power sector have an important effect on the overall composition of energy consumption.

Fuel Composition in the Electricity Sector

In the electricity sector, fuel composition is determined primarily by the technology choices when new generating capacity is built. Two factors are important for the scale of investment in generating capacity: the growth in electricity demand, and the shutdown of old power plants. Both factors are sensitive to the political and economic changes brought about in the liberalization and trade scenario.

Electricity Consumption Growth and the Requirement for New Generating Capacity

Electricity demand grows by close to 40 per cent from 1995 to 2020. However, the need for new capacity is moderate before 2005 because there is surplus capacity in all the Visegrad countries and also to some extent in Western Europe. Therefore, net capacity requirements from electricity consumption growth are only 4 GW per annum between 1995 and 2005. This rises to 7.5 GW per annum after 2005.

Plant Retirements

Throughout most of the period, the retirement of old power plants is more important for new plant construction than additional demand. Plant closure increases from 4.5 GW in the mid-1990s to 8 GW in 2000 and further to 12 GW by 2020. For the entire period from 1995 to 2020 some 60 per cent of new investments are to replace retired plants. This development is partly determined by the age profile of power plants. Construction of new power plants was mainly undertaken from 1960 to 1985; many of these plants approach the end of their economic life soon after the turn of the century.

Also important are electricity-sector reforms and the liberalization of fuel markets. In the short and medium term, the dismantling of coal subsidies has an effect, and technologies based on natural gas are the preferred alternative in

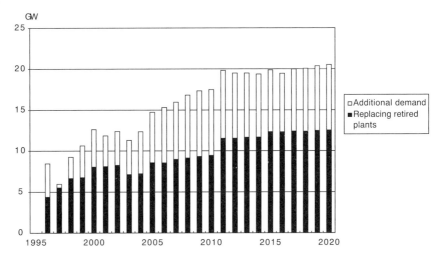

Fɪɢ. 11.6. Investment in new generating capacity (annual gross additions)

a liberalized market. In the longer term, the retirement of nuclear capacity comes to the forefront and natural gas is no longer the only alternative for new capacity; advanced coal technology becomes more attractive. In addition new renewable energy sources are making headway in electricity generation, through direct governmental support, but also in response to the new regulatory regime and institutional structure of the electricity sector. These new developments in composition of generating capacity and fuel demand will be examined below.

Dismantling Coal Subsidies—a ' Dash for Gas'?

To judge from the experience from the United Kingdom, liberalization of fuel markets and the electricity sector would result in a massive shift from coal to natural gas. During a few years in the early 1990s the British electricity-supply industry ordered more than 9 GW of new gas-fired capacity, with the potential to displace nearly 30 million tonnes of coal. From 1992 to 1995 the electricity industry's purchase of domestic coal fell from 65 million tonnes to less than 30 million tonnes. The radical changes in the UK are not duplicated in continental Europe, for several reasons.

First, in continental Europe the liberalization of the electricity sector tends to be relatively slow compared to the swift reform undertaken in the United Kingdom. Second, the oligopolistic structure of the British electricity-supply industry, with relatively high prices set by the two dominant generators, spurred a wave of construction of combined-cycle gas plants by independent power producers, with an ensuing overcapacity in power generation and premature scrapping of coal-fired plants in the late 1990s. This failure to install competition in power generation is not repeated in continental Europe. Third, the targets for reductions of SO_2 and NO_x gave further incentives to switch from coal to gas in the UK electricity-supply industry (see Newbery, 1994). The combination of environmental targets and sector reforms does not have effects of the same order of magnitude in other countries.

Still, the reforms in energy sectors in continental Europe lead to significant fuel substitution towards gas. The factors behind this development are complex and do not unequivocally work in the direction of natural gas. Effects are greatest in countries with major support programmes for national coal industries which are eliminated in parallel with the electricity-sector reforms: i.e. Germany, Spain, and the Central European countries.

In Germany and Spain, coal-sector support has been financed partly by high coal prices and partly by direct financial transfers from the state, although the High Court in Germany ruled the 'Kohlepfennig' unconstitutional in 1994 and it was eventually eliminated in 1996, see section 2.2 above. Coal prices to utilities are therefore kept substantially above world-market prices; see Fig. 11.7. In Central Europe, the support is generally provided through direct

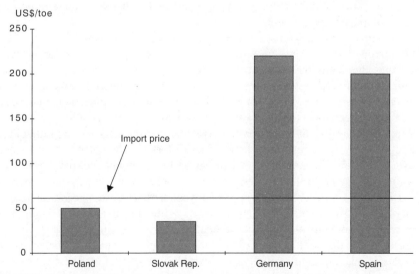

FIG. 11.7. Steam coal prices for electricity generation in selected European countries, 1994

financial transfers from the state; coal prices paid by power producers have tended to be below international prices, although prices were increased substantially from 1990 to 1994. Elimination of coal subsidies implies, therefore, that coal prices for electricity generation are radically reduced in Germany and Spain, while they increase slightly in Central Europe. Intuitively we would expect coal consumption to increase markedly in Germany and Spain as coal prices drop and to fall somewhat in Central Europe in response to higher coal prices.

The effects in Central Europe, as steam coal prices increase, are in line with this proposition; coal consumption goes down as higher coal prices trigger fuel substitution away from coal (*substitution effect*), and higher fuel costs lead to lower electricity consumption and a further reduction in coal demand (*scale effect*).

In Western Europe the price effect works in the direction of higher coal consumption. The most marked impacts on electricity prices and thus electricity consumption are found in Germany. By the year 2000 electricity consumption is 8 per cent higher than it would have been if coal prices had remained constant. This major increase follows as a result of the considerable reduction in electricity-generating costs: power plants can purchase coal at less than half the price prior to the elimination of subsidies. In Spain the impact is somewhat less since coal has a smaller share in total electricity generation (40 per cent versus 57 per cent in Germany), and coal is therefore a less important cost component in total electricity supplies.

The direction of the substitution effect in Germany and Spain is less evident. On the one hand, lower coal prices improve the competitive position of coal. On the other hand, the elimination of coal-sector support implies the lifting of purchase obligations for domestic coal that had previously been imposed on electricity generators. They are now given the option of other fuels than coal; and for power producers coal was not their economic choice in the first place. Therefore the substitution effect can give coal a lower market share rather than a higher one. The strength of these forces differs from country to country and over time, depending among other things on the timing of fuel-market liberalization compared to the progress of electricity-sector reforms.

Decisions on decommissioning and investments in power-generating capacity are highly sensitive to fuel prices in a liberalized market. Higher coal prices encourage dismantling of old coal plants, whereas lower prices work in the opposite direction. Power plants in Germany and Spain located close to domestic coal mines are initially faced with higher prices from these sources as the support to coal production is eliminated. Such power plants are generally old and it is not economically justifiable to have them refurbished to run on imported coal. Their location would also require investment in coal-transport facilities. Therefore some plants are closed earlier than they would have been under a regime with purchase obligations for coal and lack of competition in power generation. On the other hand, free access to imports of coal at low prices implies that some plants have their commercial situation improved, and this defers the retirement of plants.

In the medium and long term, however, the dismantling of coal subsidies means that electricity generators choose CCGT rather than coal technologies when the old plants are replaced by new ones. Thus the substitution effect works in the direction of lower coal demand, and this effect is more important than the higher coal demand that comes in the short term as a result of lower coal and electricity prices.

The Future of Nuclear Power

Without governmental support and control, nuclear could never have risen to account for one-third of European electricity generation by 1995. The commercial and other risks related to investments in nuclear power plants continue to require state guarantees and support. Further growth in nuclear power capacity therefore hinges on active support from public authorities in several areas, including streamlining licensing procedures, waste disposal, decommissioning, and ultimately state guarantees in case of major accidents and technical problems that can disrupt production and thus revenues.

However, it is not consistent with the priorities in economic and energy policy in this scenario to expose the public sector further to the risks of nuclear production, supporting it at the expense of other generating technologies. Construction of nuclear plants therefore comes to a standstill. Some new plants are made operational in France, but by 2005 the building of new plants has come to a complete stop. Seen in a fifty-year perspective, from 1970 to 2020, the construction of nuclear plants in Europe is highly concentrated in time with most plants coming into operation in late 1970s and early 1980s (see Fig. 11.8).

There were some, but relatively few, plant closures prior to 1995. Beyond 1995 the decommissioning of nuclear plants is affected by the shift in policy and the new business environment. Although most nuclear plants are kept in public ownership after the electricity sector reforms, investments and operational dispositions are made primarily on the basis of commercial considerations. The costs of keeping the old plants in operation are increasing, given the continued stringent safety standards. Hence many of the plants are, according to the stipulated economic lifetime schedule, to be decommissioned between 2010 and 2020. Major investments are undertaken to extend the lifetime of

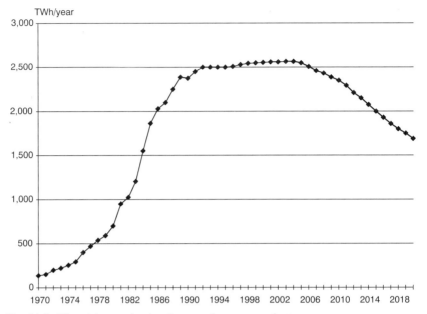

Fɪɢ. 11.8. Electricity production from nuclear power plants

some of these plants, and such investment often yields a good rate of return. For other plants investments are not justifiable on economic grounds. In such cases the alternatives are either to provide financial transfers from state budgets, or to close down the plants. The latter is often the preferred option, given the technical condition of the plants in question.

Decommissioning of nuclear plants therefore has a profound impact on the structure of electricity generation after 2010. By then many plants have operated for more than thirty years. It is normally assumed that the maximum technical lifetime of nuclear plants is forty-five years. By 2020 only 10 per cent of the capacity in operation in 1995 will be older than forty-five years.

Between 1995 and 2010 less than 5 GW of nuclear capacity are closed down, primarily in the United Kingdom and some old plants in Germany and France. If no decommissioning had taken place between 1995 and 2010, the average age of the plants in 2010 would have been thirty years, and a quarter of the nuclear capacity would by then be older than thirty-five years. Since there is no new nuclear capacity coming into operation after 2005 the average age of the plant stock increases by one year every year, which mean that by 2010 a large part of the plants are nearing the end of their normal economic life.

On average, nuclear plants are retired at an age of thirty-five years up until

2010. After 2010 policies are stepped up to keep more plants in operation beyond what can be justified on purely economic grounds. The underlying rationale is the increasing reliance on energy imports. Towards 2020, therefore, the average age of plants being retired gradually increases to thirty-eight years. This cannot prevent an acceleration in plant closure, which inescapably follows as a result of the age profile of nuclear power plants, again caused by the rapid growth and fall of nuclear construction back in the 1970s and 1980s.

Combined-cycle Gas versus Advanced Coal Technology

As explained above, both electricity-sector reforms and the dismantling of coal subsidies spur investments in modern gas-fired technologies. From 1995 to 2005 the increase in natural gas use in electricity generation is 35 per cent higher than the total net increase in primary-energy use for electricity generation, i.e. natural gas is the principal choice for meeting net demand and for replacing retired capacity. The cost advantage of combined cycle gas during this period was discussed in section 10.4. After 2005 this picture starts to change as fuel prices move steadily in favour of coal, and advanced coal technology comes into commercial use. Whereas the 1980s and 1990s saw major improvements in gas-fired technologies, improvements in new coal technologies follow after the turn of the century, as a result of major R&D efforts in light of the favourable price prospects for coal and the continuous increase in power-generating capacity worldwide. However, in a liberalized electricity market, the lower up-front capital costs of combined cycle gas plants implies that this technology will remain competitive with advanced coal technologies despite higher prices for natural gas. Consumption of both coal and natural gas grows rapidly after 2010, not least because of the decline in nuclear power production.

Renewable Energy Resources

It is not only natural gas and coal that gain market shares: renewable energy sources also show a steady growth. From 1995 to 2020 electricity production from renewable energy sources increases from 460 to 710 TWh/year. Its share of total electricity generation rises slightly, from 20 to 22 per cent. Hydropower is the principal primary source of renewable energy for electricity generation. However the growth of hydroelectric power generation is moderate compared to other categories of renewable sources; see Fig. 11.10. This is a result of the environmental opposition to the development of the remaining— albeit modest—economic potential of hydropower. Moreover, there is a major policy push for new renewable energy sources. By 2020, hydropower accounts for only 68 per cent of renewable electricity generation, against a share of 93

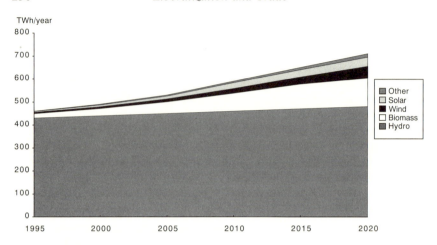

TWh/year

Fig. 11.9. Electricity generation from renewable energy sources

Note: *Renewables' share of electricity generation is higher than their share of primary-energy input shown in the figure. The reason is that energy input for electricity generation by renewables, in accordance with international statistical conventions, is accounted as having the same heat value as the output of electricity, whereas a conversion of 33% is assumed for nuclear, and the actual conversion losses are applied for power generated from coal, oil, and natural gas.

per cent in 1995. Biomass attains a share of nearly 20 per cent primarily as the result of major increases in power production from municipal and industrial waste, as well as agricultural production and residues. Wind power and solar energy also show large increases, in both cases a ten-fold increase, producing respectively 50 and 40 TWh of electricity in 2020.

These increases represent a substantial policy effort by the European Union and national governments. This support takes the form of increased allocation of research, development, and demonstration funding for renewables, partly at the expense of nuclear and to some extent also coal research. The decline in nuclear and coal research follows as a result of their decreasing role in the European energy balance. In addition, there is direct financial support and regulatory changes that promote the production of electricity from renewable energy sources.

There are several political motives for the increased attention paid to renewable energy sources. As mentioned, the concern for reliance on energy imports is growing and the scope for direct policy interventions to determine fuel choices is limited. Renewables represent one area where it is natural for the European Commission to take on an active role in response to energy-security concerns. Promoting renewables calls for coordination of national efforts: the

Box 11.2 *New renewable energy sources*

In a European context and within the time perspective to 2020, biomass/ waste, wind power, and solar are the most important new renewable sources for production of electricity and heat.

Biomass and waste
Municipal and industrial waste has a large growth potential for electricity and heat generation. In addition to the burning of solid waste, methane from landfills can be used in nearby industries or for power production. Approximately 9 TWh of electricity was generated from landfills in 1995. Although waste is difficult and costly to handle for energy conversion, stringent waste management regulations mean that waste for energy use is an increasingly attractive option in economic terms. Energy plantations can provide biomass both for the production of liquids (e.g. alcohol from sugar and wheat, diesel/fuel oil substitutes from rapeseed and sunflowers) and solids (wood). The interest in wood plantations has grown in recent years and several demonstration projects have been launched, primarily in Northern Europe. Production of liquids has been more limited but holds a large potential in consideration of Europe's overproduction in agriculture. Though not generally considered a new renewable energy source, agricultural and forest residues are the most important renewables after hydroelectric power. They are used extensively in industry and households in some countries. The potential for increased use is large but constrained both by ecological and economic factors. Removing some of the residues may yield ecological benefits, whereas removing all may be damaging. The cost of collecting residues is often high and represents an important economic obstacle.

Wind power
Wind power is commercially viable in some areas along the Atlantic and the North Sea coasts. Remote mountainous locations with difficult access to centralized power systems can also attract commercial interest. For the most part, however, the cost of wind power remains above that of centralized systems based on new thermal technologies. Even with continued cost reductions for the manufacture of wind turbines and increasing fuel prices, wind energy will in 2020 be 30–80 per cent above costs of thermal electricity generation. Therefore, the expansion in wind-power capacity must rely on public support. European electricity generation from renewables was 28,000 GWh in 1995, of which wind power

and power generation from waste contributed the largest part. The growth was rapid during the 1990s with considerable operational experience being gained. There is a large technical potential for wind energy in Europe. According to Grubb (1995) up to 10 per cent can in the long term be supplied from wind turbines.

Solar

Solar energy consists of three main categories: solar thermal, photovoltaic (PV), and thermochemical and photochemical systems (mainly for hydrogen generation). Within the first category passive solar gains in buildings have a considerable potential, particularly in Northern Europe because of the long heating season. In Southern Europe active solar water heating has a growth potential. Solar thermal power conversion is of less importance throughout the most of Europe due to lack of sustained and intense direct sunlight. Photovoltaic solar electricity can compete with diesel generation and grid extension for small-scale remote applications. Although technological innovations and large cost-reductions increase the attractiveness of PV for centralized power production, their costs will throughout the period to 2020 remain more than twice as high as power from conventional sources. Therefore, PV is in the short and medium term of less importance in Europe, with its well-developed electricity network, than in many other parts of the world. Thermochemical and photochemical systems are insignificant within the timeframe of the year 2020.

Sources: World Energy Council (1994), Grubb (1995)

European Commission, through its R&D programmes (notably ALTENER), plays an important role in this regard. Promoting renewables also fits well into EU's policy on technology and environmental concern. A solid industrial base in renewable technologies, in which European industry historically has a strong position, is perceived to give Europe a competitive advantage as global demand for alternatives to coal, oil, and gas rises.

Further benefits from promoting renewables are achieved through the alleviation of social and regional pressures that arise from liberalization and trade. Increased demand for biomass helps to cushion the consequences of liberalization in trade for agricultural products. Moreover, some lignite-based district heating systems in Central Europe are 'saved' through conversion to biomass.

The economically most attractive supply-schemes based on new renewable energy sources are small-scale and thus well accommodated into the new structure of the electricity sector. The new structure of the electricity sector provides a much improved basis for small 'independent' power producers to compete on the electricity market compared to the old system, where large utilities often held exclusive supply rights. Preferential tariffs for the supply of power from renewables, similar to the non-fossil-fuel obligation introduced in the United Kingdom in the early 1990s, become an important means of promoting renewable technologies throughout Europe.

Fuel mix

Fig. 11.10 summarizes the development in the fuel mix in electricity generation. Compared to the abrupt change from 1973 to 1985 the development from 1995 to 2020 appears smooth. But in view of the moderate growth in electricity demand (and net addition in capacity) the changes in relative shares of different technologies are quite substantial: the nuclear share falls from 40 per cent in 1995 to 22 per cent in 2020 and the natural gas share increases from 5 to 27 per cent.

Both electricity-sector reforms and the dismantling of coal subsidies spur investments in modern gas-fired technologies. From 1995 to 2005, most new generating capacity is in gas-fired combined-cycle plants. In this period gas technology enjoys a clear cost advantage over conventional coal technologies. After 2005 this picture starts to change. Fuel prices move steadily in favour of

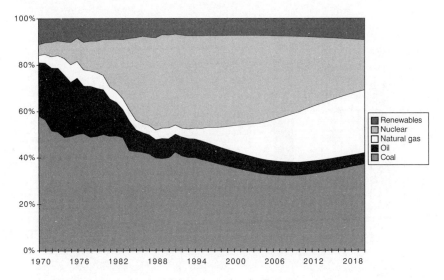

FIG. 11.10. Fuel shares in electricity generation in Europe

coal; international steam coal prices continue to decline, whereas natural gas prices increase; see Box 10.2. More importantly, advanced coal technology is coming into commercial use, making progress at the expense of combined-cycle gas. Whereas the 1980s and 1990s saw major improvements in gas-fired technologies, improvements in new coal technologies follow after the turn of the century. This is the result of major R&D efforts, in light of the favourable price prospects for coal and the continuous increase in power-generating capacity worldwide.

In general, the shifts in fuel shares from 1995 to 2020 are much larger in the electricity sector than in other applications of primary energy use; see Fig. 11.11. As noted above, the electricity sector exhibits a shift in the configuration of generating capacity which is as radical as the one in 1970–95, whereas other sectors show stability, with a modest increase in the use of gas at the expense of coal. During the first ten to fifteen years there is strong growth in natural gas use—in industry, households, and electricity generation. After 2010 this growth levels off somewhat due to higher prices for natural gas. Structural changes in industry reduce demand for all primary fuels, and most notably for coal. In Central Europe coal demand is drastically reduced in district heating, as such plants are modernized and partly converted to natural gas or are closed down. In the electricity sector, coal loses ground to natural gas once the protection of domestic coal production is lifted.

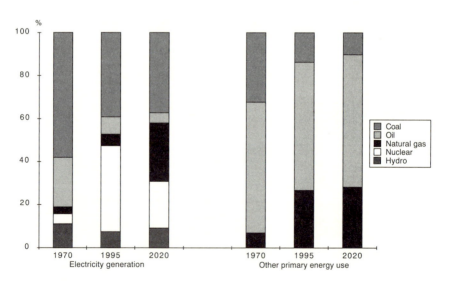

FIG. 11.11. Energy consumption by primary fuel in Europe

Oil continues to lose market shares in industrial processes and for space heating in industry, services, and households. However, the relatively strong increase in transport-fuel demand implies that oil maintains its market share of about 40 per cent throughout the period.

Energy Imports

The shift in the composition of energy demand has substantial effects on energy imports; see Fig. 11.12. From 1995 to 2020 the import share of energy consumption increases from 39 to 57 per cent. Stagnant or falling production of primary energy in most European countries also contributes to this trend.

Taken separately, all three fossil fuels show marked increases in import shares. This means that European production is unable to keep pace with the development in demand. In the case of coal, this is the result of the large reduction in coal mining primarily in Germany and in Central Europe. For oil it follows from reduced European production and a 25 per cent increase in oil consumption from 1995 to 2020. The relative increase in import dependence for natural gas is somewhat lower despite rising gas demand to 2020. Higher production in Norway and in the United Kingdom from 1995 to 2005 helps to curb the growth in imports. Increases in natural-gas prices from about 2005 also help to extend the production increases of natural gas from the North Sea which otherwise would have levelled off after 2005. After 2010, however, some 80 per cent of incremental energy demand is met by imports.

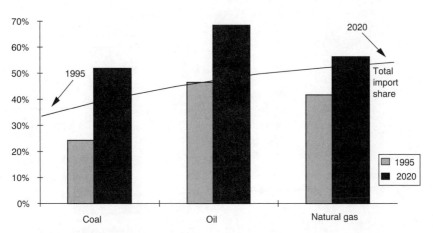

FIG. 11.12. Import shares for coal, oil, natural gas, and total primary energy in Europe, 1995–2020

The perceived commercial risks of linking up to more gas use help to dampen the growth in demand. This effect is accentuated by the reforms of the natural-gas markets with more competition in the market and ensuing volatility in gas prices. To some extent, therefore, the market 'internalizes' the risk of a supply disruption to gas supplies and the economic damage that might follow. But despite considerable policy efforts, reliance on energy imports from potentially unstable suppliers increases substantially.

12
Liberalization versus National Rebound

This chapter summarizes and compares energy trends and associated emissions of CO_2, SO_2, and NO_x in the two scenarios. As described in the two preceding chapters, the institutional structures and regulatory systems of the two scenarios differ. This has major implications for the composition of energy demand. However GDP growth rates or energy-price trends, two main determinants for energy-consumption trends, differ relatively little between the scenarios. Economic activity and energy prices are however not identical—but the differences result from sector reforms and different trade regimes and are not caused by 'good or bad' macro-economic policy or differences in international energy price trends. Nor do the scenarios assume differences in the political focus on energy conservation. Thus there are no 'vigorous efficiency efforts' to generate diverting energy trends, as so often assumed in scenario analyses.

Differences in energy and emission trends presented in this chapter show the effects of different-energy sector structures and broad political and economic reform, rather than the impacts of an explicit policy to curb energy demand and associated emissions.

12.1. Energy Trends

Total Energy Consumption

Comparing the differences in energy consumption between the two scenarios, we can see disparate trends in Central and in Western Europe. In Western Europe energy consumption is constantly lower in liberalization and trade. The difference widens slowly to 2 per cent of total primary energy consumption by the year 2020. This relatively modest difference reflects offsetting factors that influence energy demand in the two scenarios. In Central Europe the difference in consumption level varies significantly over time. From 1995 to 2005 energy

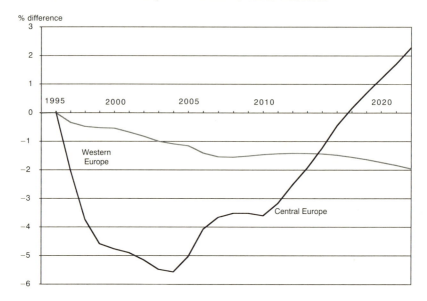

FIG. 12.1. Total energy consumption in liberalization and trade relative to national rebound (% difference)

consumption in liberalization and trade averages 4 per cent below the level in national rebound. After 2005, energy consumption growth in liberalization and trade is persistently higher than in national rebound; and after 2015 it outstrips the level of national rebound.

Energy Consumption by Sector

Central Europe

Differences in primary energy use by sector are shown in Fig. 12.2. The category *direct stationary consumption* (direct use of coal, oil, and natural gas in industry, service sectors, and households) shows a much weaker development from 1995 to 2010 in liberalization and trade. One reason for this is modernization of industry and district heating plants, resulting both in fuel switching and in energy-efficiency improvements. Industrial restructuring and the effects of energy-price reforms are the principal underlying factors. After 2010, however, direct stationary energy consumption grows more in liberalization and trade, because the largest and most obvious energy-efficiency

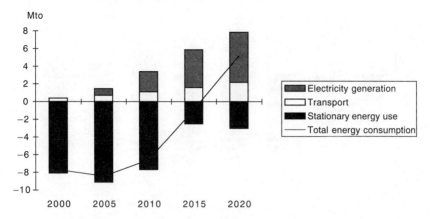

Fɪɢ. 12.2. Difference in primary energy use by sector in the two scenarios, Central Europe

potential has now been exploited, and because of continued higher economic growth.

Transport fuel consumption is constantly higher in liberalization and trade, due to higher disposable income and the growth in demand for freight transport. However, the differences are relatively small (transport-fuel consumption grows rapidly in both scenarios) because of higher fuel efficiency in liberalization and trade. Both electricity demand and fuel demand for electricity generation are slightly higher during the first five to six years in national rebound. This is caused by wastage in electricity end-use and in the electricity-supply industry. Substantial price increases for households, as well as investment in modern technologies, keep growth in check until the turn of the century. Thereafter high economic growth triggers a rapid growth in electricity demand which contributes to substantially higher fuel use in liberalization and trade.

Western Europe

Stationary energy use is lower in liberalization and trade despite higher economic growth. There are two factors that result in the lower consumption. First, industrial restructuring reduces energy demand and stimulates the substitution from direct fuel use to electricity. Second, large market shares for natural gas contribute to lower energy intensities because gas combustion normally has higher technical efficiency than oil and coal.

Transport fuel consumption is higher in liberalization and trade, for the same reasons as in Central Europe. However, the relative percentage difference is

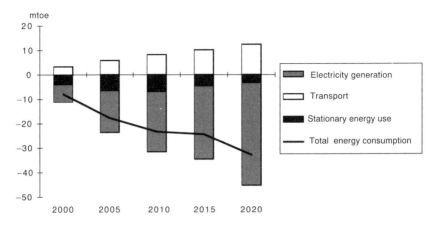

Fɪɢ. 12.3. Difference in primary energy use by sector in the two scenarios, Western Europe

smaller in Western Europe, for two reasons: the difference in GDP between the two scenarios is lower in Western Europe, and the demand for transport fuel responds less to economic growth. Moreover, the share of rail transport in freight of goods is increasing. Western European countries have substantially lower levels of power-sector fuel consumption in liberalization and trade than is the case in Central Europe. One reason is that electricity demand responds less to economic growth in Western Europe. Another is that industrial restructuring results in slightly lower growth in electricity demand. And, most importantly, changes in the fuel mix from coal and nuclear to natural gas yield much higher thermal efficiency in liberalization and trade. In total, therefore, primary fuel use for power production in 2020 is about 7 per cent lower in liberalization and trade.

Fuel Mix

The fuel mix shows notable differences in both Central and Western Europe. Natural gas and renewables have higher shares in liberalization and trade, at the expense of coal and nuclear. The electricity sector, transport, and other sectors all contribute to this pattern. From 1995 to 2010 there is a decline in the coal share in both scenarios, for Western and Central Europe alike. In Central Europe the relative decline of coal continues after 2010, but in Western Europe the development is different. After 2010 coal consumption in liberalization and trade increases quite rapidly. By 2020, coal for power generation is actually higher in liberalization and trade.

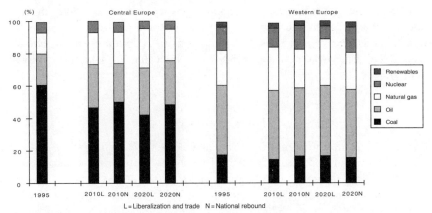

FIG. 12.4. Total energy consumption by fuel, 1995–2020

Natural gas consumption increases by 75 per cent from 1995 to 2020 in liberalization and trade, as against 50 per cent in national rebound. It is primarily the continued growth in nuclear power generation that curbs the growth in natural gas in national rebound. Higher coal use in industry and district heating also has an impact. Nuclear power is the primary energy source with the most significant difference in fuel shares. In liberalization and trade, its share falls from 14 to 7 per cent, whereas it stays constant in national rebound. Western Europe is much more affected by nuclear decommissioning and lower production regularity in the liberalization scenario than Central Europe.

Oil consumption is by 2020 4.5 per cent higher in liberalization and trade. Oil has a higher growth rate in all sectors except in electricity generation, where more of the old capacity is being retired. Transport consumption is higher, as shown in Figs. 12.2 and 12.3. Oil for heating purposes also shows lower growth in national rebound because of less substitution from coal. Renewable energy sources show a weaker development in national rebound. Investment and operational support to these technologies is lower, both from EU funds and from national governments. More energy-sector support, including R&D, is allocated to coal and nuclear.

12.2. Environmental indicators

Overall energy-consumption trends may differ by relatively small margins, but this is not the case for emissions of air pollutants and CO_2. Both fuel

switching and emission control technologies contribute to more distinct
emission paths. Local and regional air pollutants and CO_2 move in different
directions in the two scenarios. CO_2 emissions in 2020 are higher in liberalization and trade, whereas SO_2 and NO_x, as indicators of local and regional
damage, are lower.

Emissions of CO_2

CO_2 emissions in Europe increase by nearly 26 per cent from 1995 to 2020 in
liberalization and trade; see Fig. 12.5. This rise is caused by two factors: the
overall increase in energy consumption, and a shift in energy demand towards
higher carbon intensity. Carbon intensity in liberalization and trade shows a
continuous decline from 1970 to 2005. Initially this decline is caused by oil
and gas replacing coal in industry and other sectors, and nuclear replacing
fossil fuels in the electricity sector. This trend was prolonged after 1995 by the
dismantling of coal subsidies in Western and Central Europe, and further

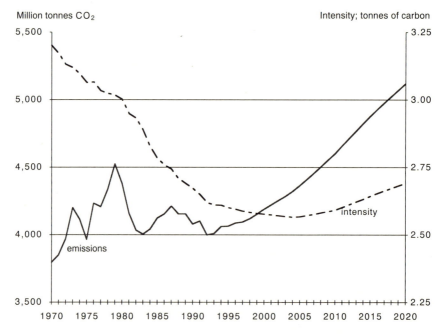

FIG. 12.5. CO_2 emissions and carbon intensity in the liberalization and trade scenario
in Europe

induced by economic restructuring. However, the tightening of the natural gas market and the phase-out of nuclear plants contribute to a reversal and even an upward trend in carbon intensity from 2005. In national rebound, carbon intensity levels off earlier but then stays constant at about 2.6 tonnes CO_2 per toe energy until 2020. After 2010, carbon intensity is lower in national rebound.

By the year 2020 CO_2 emissions in liberalization and trade are 5.5 per cent higher than in national rebound. However, before 2010 emissions are higher in national rebound in both Western and Central Europe. This reflects higher overall energy consumption, and coal consumption in particular, during the first fifteen to twenty years of national rebound. Towards the end of the period, the difference in energy consumption in the two scenarios becomes small, and coal consumption in liberalization and trade starts to increase.

The aggregate emission trends, however, mask a set of opposite developments, which are illustrated in Figs. 12.7 and 12.8. Fig. 12.7. shows that the emissions level in Central Europe is initially much lower in liberalization and trade. This is because direct energy consumption in stationary use, and in particular in coal consumption, is lower than in national rebound. This effect eases off somewhat after 2010 as consumption growth in liberalization and

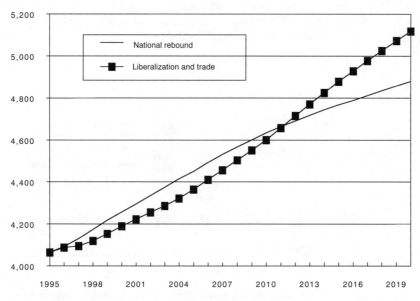

FIG. 12.6. CO_2 emissions in two scenarios, 1995 and 2020

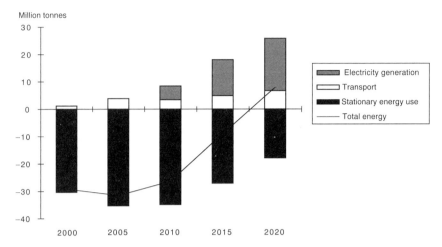

Fɪɢ. 12.7. Difference in CO_2 emissions by sector in the liberalization and trade scenario in Central Europe

trade accelerates. After 2005, high growth in electricity demand contributes to higher emissions in liberalization and trade. However, the effect on CO_2 emissions is less than for primary energy use (see Fig. 12.2.) because natural gas has a higher share in liberalization and trade. Additional emissions from transport in liberalization and trade are modest.

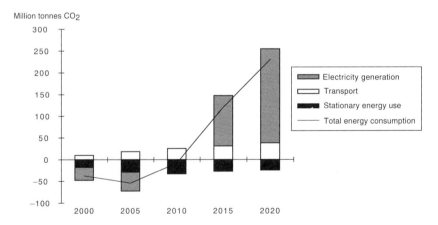

Fɪɢ. 12.8. Difference in CO_2 emissions by sector in the two scenarios in Western Europe

In Western Europe emissions before 2010 are low in liberalization and trade. Both direct stationary energy use and electricity generation contribute to this, because of lower fuel demand and a fuel mix with lower carbon intensity. After 2010, however, this changes drastically. Growth in CO_2 emissions is much higher in liberalization and trade, whereas fuel consumption growth is lower, as explained above. This is caused by the phase out of nuclear capacity in liberalization and trade. Keeping electricity out of the equation, emissions from other sectors are in aggregate roughly the same in the two scenarios, i.e. higher emissions from the transport sector are counterbalanced by lower emissions from direct stationary use.

Emissions of SO_2 and NO_x

As explained in Chapter 7, the emissions of SO_2 and NO_x relative to economic activity and energy consumption were much higher in Central Europe than in Western Europe throughout the 1980s. These differences narrow from 1995 to 2020, but are still significant in 2020. Moreover, the emission trends are more distinct than is the case for CO_2.

Western Europe

The downward trend in SO_2 emissions which started around 1980 continues after 1995, though at a slower pace. By 2020 emissions are less than half the 1995 level and only a quarter of the 1980 level. From 1995 to 2005 the major part of the decline is in Germany and the United Kingdom. Later there are notable reductions in Spain and Italy, where the decline is modest during the first part of the period. The reduction in 1995–2005 comes largely as a result of fuel substitution towards products with low sulphur content. In the UK, natural gas replaces much indigenous coal for power generation. In Germany, lower lignite consumption and installation of emission control technologies in the new Länder are important factors. The latter is also an important reason for the lower emission trajectory in liberalization and trade; see Fig. 12.9. Industrial restructuring also contributes to lower SO_2 emissions in liberalization and trade prior to 2005. After that time, the reduction runs more or less parallel in the two scenarios, as the phase out of nuclear plants contributes to higher coal consumption in liberalization and trade.

Emissions of NO_x show a 15 per cent decrease from 1995 to 2020, partly attributable to natural gas replacing coal in power plants and industry. However, this reduction is more modest than is the case of SO_2, for two reasons. First, consumption of transport fuels, which contribute more than half of NO_x emissions, continues to grow, offsetting some of the emission reductions in

Million tonnes

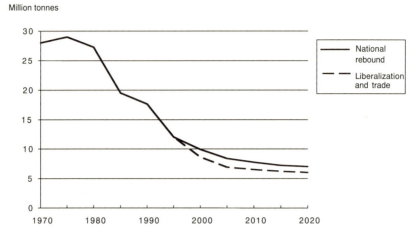

FIG. 12.9. Emissions of SO_2 in Western Europe

other sectors. Second, emission controls for NO_x are more costly and technically more difficult to achieve than for SO_2. The large role of transport fuel in
NO_x emissions and the relatively modest role of fuel switching for emissions
reduction indicates relatively small differences in emission paths between the
two scenarios. The transport sector contributes to higher emissions in liberalization and trade, but this is more than offset by higher energy-efficiency
improvements and lower NO_x emissions from fuel composition in other sectors
of energy use. The power sector has lower NO_x emissions in liberalization and

Million tonnes

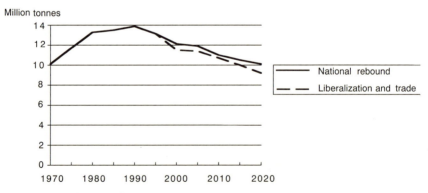

FIG. 12.10. Emissions of NO_x in Western Europe

trade all through the period to 2020. This contrasts with CO_2 emissions, which are higher in liberalization and trade; see Fig. 12.8.

NO_x emissions are lower in power generation in liberalization and trade because the high share of new generating capacity prior to 2010 is gas-based, whereas conventional coal has a larger share in national rebound. When coal-fired power increases its market share in liberalization and trade after 2010, this is in the form of advanced coal combustion with NO_x emissions much lower than for conventional coal combustion.

Central Europe

The reduction in Central Europe's SO_2 emissions, which accelerated when the transition to a market economy started in 1989, continues after 1995. Whereas the reduction in emissions prior to 1995 resulted largely from economic recession and close-down of the worst polluting power and industrial plants, later emission reductions are increasingly the result of installation of air-pollution controls, fuel switching, and energy-efficiency improvements. These factors are much stronger in liberalization and trade, yielding substantial differences in emission levels between the two scenarios—ranging from 15 to 20 per cent. Even when energy demand in liberalization and trade exceeds the level of national rebound after 2015, there is a significant gap in emission levels between the scenarios.

Emissions of NO_x are also declining, but less than for SO_2. From 1995 to 2005 we can see an increasing gap between the emission levels of the two scenarios. Emissions are lower in liberalization and trade for the same reasons as noted for SO_2; however, the relative difference is smaller for NO_x and it narrows to only 15 per cent by 2020. The reason for this is higher energy

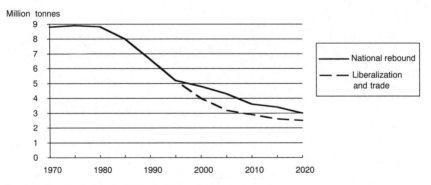

FIG. 12.11. Emissions of SO_2 in Central Europe

Million tonnes

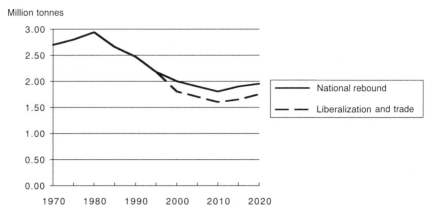

FIG. 12.12. Emissions of NO$_x$ in Central Europe

consumption and in particular transport fuel consumption in liberalization and trade. The higher proportion of end-of-pipe technologies being applied in liberalization and trade prevents NO$_x$ emissions from exceeding the emission level in national rebound.

13
Environmental Futures

In this chapter, we analyse the scope for and effect of a very ambitious environmental policy imposed on the energy structures in Europe as depicted in two scenarios: national rebound, and liberalization and trade. We start from the assumption that climate change and local and regional air pollution have the highest priority on the policy agenda.

Both the scenarios presented in the preceding chapters encompass a policy of moderate environmental concern. In this chapter we shall examine the policies and the energy and environmental impacts that follow as the result of a shift from moderate to high environmental concern. This radical increase in environmental concerns is expressed as new and ambitious targets for emission of CO_2, SO_2, and NO_x. To meet these targets, aggressive new policies are developed and implemented. The selection of policy instruments is highly dependent on the scenario in question. For institutional and political reasons, the logic embedded in liberalization and trade predisposes to other responses to the environmental challenges than in the case of national rebound.

13.1. Emission Targets and Abatement Costs

Targets for NO_x and SO_2

The 1994 Oslo Protocol on SO_2 sets a reduction target in European emissions of 40 per cent by 2010 compared to the 1995 level. We assume that during the second part of the 1990s a new and strengthened protocol is signed that requires a reduction of 60 per cent by 2010 and 70 per cent by 2020. This implies that European SO_2 emissions in 2020 will be at less than 15 per cent of the peak in 1975. The negotiated emission reductions vary from country to country, with costs of reduction and environmental impact playing an important role. Central and South European countries are faced with the largest reduction targets.

For NO_x, the new protocol is hypothesized to contain a 20 per cent reduction

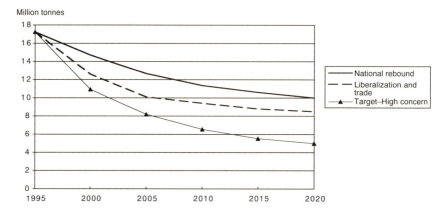

FIG. 13.1. Emission targets for SO_2

by the year 2010, compared to the 1995 level. High environmental concern results in negotiated reduction of a further twenty percentage points by 2010. For 2020, targets are set at 40 per cent of the 1995 level. The percentage reductions for individual countries vary, but less than for SO_2 since few countries achieved major reductions in NO_x emissions prior to 1995. Emission reductions for SO_2 and NO_x in accordance with these targets bring emissions to a level 'which according to present knowledge do not bring harmful effects to the environment' (critical loads).

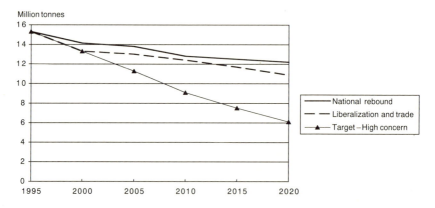

FIG. 13.2. Emission targets for NO_x

Targets and Abatement Costs for CO_2

For CO_2 the target level in 2020 is 20 per cent below 1995 emissions. This target is less ambitious than the statement at the 1988 Toronto Conference that urged governments to reduce emissions to 20 per cent of the 1987 level by 2005. Compared to the emission developments of moderate concern, presented in Chapter 12, this represents a reduction in 2020 of 34 per cent and 37 per cent for national rebound and liberalization and trade respectively. The targets are part of a new protocol under the Climate Change Convention with both the European Union and individual countries as signatories. In national rebound, the commitments are for each individual country to undertake measures sufficient to meet the targets within their own national borders. In liberalization and trade there is one European target, and actions to meet the target are formulated mainly at the EU level. Percentage reduction in emissions may vary substantially between countries, depending on the effects of measures imposed by EU bodies.

If EU policy in liberalization and trade were dictated only by cost efficiency (i.e. reductions in emissions are made where they entail the lowest possible costs) emission reductions would differ by country, reflecting national variations in marginal abatement costs. This is illustrated for France, Germany, and Central Europe in Fig. 13.4. The horizontal line in the Figure indicates the marginal abatement cost at which the EU target is achieved. At this level, emission reductions in France, Germany, and Central Europe would be 12, 18, and 30 per cent respectively.

In national rebound all countries are committed to the same 20 per cent

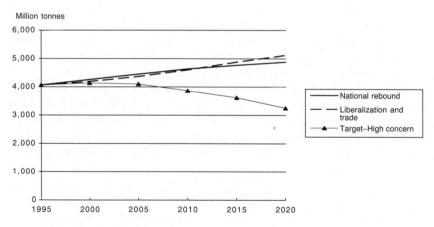

FIG. 13.3. Emission target for CO_2

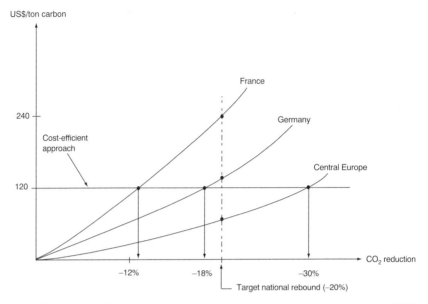

US$/ton carbon

240

Cost-efficient
approach

120

France

Germany

Central Europe

CO₂ reduction

−12%   −18%   −30%

Target national rebound (−20%)

FIG. 13.4. Marginal abatement costs and emission targets in selected countries, 2020

reduction and the marginal abatement costs differ between countries. The point where the vertical line crosses the abatement cost curves in the Figure indicates the marginal abatement costs in national rebound.

The Figure illustrates variations in country-specific emission reductions and associated abatement costs between the two scenarios. In liberalization and trade, abatement costs are relatively low if a policy is pursued that equalizes marginal costs among countries. However, this would generate sizeable differences in the total costs imposed on individual countries. Some countries would undergo major structural changes whereas others would be only slightly affected. In liberalization and trade, where the reduction target is for Europe as a whole, burden-sharing becomes an issue that influences the policy measures imposed at the EU level. In national rebound the problem of burden-sharing is 'solved' by countries agreeing to the proportional reduction from the 1995 base year. This means that marginal abatement costs show large differences by country.

Focus on Climate Policy

Climate policy is the overriding issue on the environmental policy agenda. For there to be reductions in CO_2 emissions of 20 per cent over a twenty-five-year

Box 13.1 *Marginal abatement costs*

Marginal abatement costs for CO_2 are the costs of reducing emissions by one additional unit. Normally it is assumed that the marginal costs increase with increasing emission reductions. This can be illustrated as an abatement cost curve, which shows the marginal abatement costs for any given reduction in CO_2 emissions; see Fig. 13.4.

In Central European countries, required emission reductions in 2020 are large because unabated emission growth is high. This is caused by high economic growth and continued heavy reliance on coal in electricity generation and industry. Abatement costs, however, are low because energy prices are low and the coal share in total energy demand is high. This gives ample opportunities for inexpensive energy conservation and fuel substitution to reduce emissions.

France, at the other end of the scale, has high energy prices, which indicates that incentives have been put in place to economize energy use. Further large-scale improvements are costly. Moreover, there are fewer opportunities for reducing CO_2 emissions through fuel substitution in France than elsewhere since coal has only a negligible share in electricity generation. In addition, a large part of growth in emissions stems from the transport sector where fuel switching and energy conservation is very expensive.

Germany is a case in between these two extremes. On the one hand, energy-price levels in Germany are as high as in France, a factor which points towards high abatement costs. On the other hand, Germany resembles the Central European countries in the sense that energy consumption is carbon-intensive, thus with large opportunities for emission reductions through fuel switching. This makes the costs lower than in France.

period, great political efforts will be required, with large ramifications for economic activity and individual behaviour. Since removing of CO_2 from energy combustion is not economically viable, measures will have to be found that promote energy conservation or fuel substitution. They are generally costly and exceed by a large margin the costs of installing control technologies for SO_2 and NO_x emissions. For this reason, climate policy is the focus of our analysis of environmental policies in the two scenarios. However, the impacts on NO_x and SO_2 emissions are considered during the process of designing a strategy to meet the CO_2 target. Impacts of climate policy on SO_2 and NO_x emissions are analysed towards the end of this chapter.

Box 13.2 *Policy instruments to reduce emissions*

It is common to distinguish between four categories of policy instruments to reduce emissions:

Emission controls—through technical standards on machinery and equipment, product quality standards, and direct emission standards. They have in the past widely applied in relation to SO_2, NO_x, and a range of other pollutants.

Emission charges have in the past played only a minor role in environmental policy. Sulphur taxes and tax discrimination of lead in petrol have been applied in many European countries, but the impact on total energy consumption trends have been low. Since around 1990, however, carbon taxation has received considerable attention among analysts and policy-makers. A carbon tax is in theory the most cost-efficient instrument for achieving reductions in CO_2 emissions. The potential damage of CO_2 emissions is independent of where emissions take place, and emissions are proportional to the carbon content of the fuels. It is therefore straightforward to impose a tax on each fuel, proportional to its contribution to climate change. This contrasts with the complexity of implementing efficient charges on SO_2 and NO_x. Environmental damage from these emissions depends on location and on a variety of fuel qualities. There are various technological options for controlling emissions.

Energy-conservation policy (non-price measures), which includes technical standards, information and financial-support programmes, voluntary or negotiated programmes to conserve energy, aims at improving energy efficiency and encourages energy saving over and above what is achieved through pricing and taxation policies.

Fuel substitution measures comprise direct or indirect policy interventions other than taxation and price-setting. Public authorities can affect the fuel choices in the electricity sector by influencing investment decisions for generating capacity. Investments in new power plants are controlled through licensing procedures or through direct public-sector involvement in power sector strategies and operations. Measures that accelerate the rate of plant retirement are also important, as they often replace old, inefficient, and polluting coal technologies with modern gas-fired plants with high thermal efficiency and low emissions.

13.2. Climate Policy in National Rebound

The National Framework for Policy

In national rebound, every country in Europe adopts its own policy to reduce domestic CO_2 emissions. The variations in economic and political costs of meeting this target depend on the structure and growth in energy production and consumption, and the trade-offs that arise between meeting the emission target and other pressing policy objectives. In general, a 20 per cent reduction in emissions of CO_2 is a very ambitious target for all countries involved, one which will require new and radical policies to be introduced and kept in place over several decades. As a result, the energy-sector structures of this scenario are reinforced and the framework for policy remains in place. Strategies to reduce emissions are worked out in close cooperation with the dominant actors in the energy sector, in particular the existing and monopolized utilities in the electricity market. Hence a new social contract emphasizing environmental concerns is established between the energy industries and the public.

Policy options are identified within the domain of the nation-state and measures are selected that minimize side-effects detrimental to domestic interests. However, the radical nature of the emission targets calls for policies that go far beyond what has been implemented in the past, policies that will have large economic and social impact. During the first phase of climate policy formulation, measures are chosen similar to those that were effective in reaching the policy objectives of the 1970s and the 1980s (see Chapters 2–6). Emphasis at that time was on interventions that changed the fuel mix in the electricity sector, energy conservation through administrative measures such as DSM-planning in utilities, information campaigns, and tightened building codes. But such measures, even if pursued aggressively, cannot ensure that the emission targets are reached. Therefore, a variety of new policy instruments are called for. Important in this regard are taxes and levies on energy use.

The design of tax systems does not resemble the textbook prescription of a uniform carbon tax across fuels, sectors, and countries. The reason for this is the national confinement of taxation policies, which means that consideration has to be given to the competitive position for exposed industries, including the power sector, and the interests of domestic producers of primary energy, primarily coal production. Therefore taxes (for political reasons labelled carbon taxes) are levied rather selectively on households and some other domestic end-users, in addition to imported fuels like oil and increasingly natural gas.

Carbon Taxes

If a carbon tax were imposed as the only climate policy measure, increases in energy prices would be radical. Price increases for industry would be particularly large. In France, for example, natural-gas prices for industry would increase by more than 200 per cent and the impact on industrial competitiveness would be severe. Sectors exposed to international competition are therefore either exempted from the tax or are granted a lower tax rate. By and large, energy-intensive industries enjoy full exemption from the carbon tax. Reductions in emissions from these sectors are primarily pursued through regulatory measures and voluntary agreements between governments and industries. Energy use in refineries and in aviation is not taxed.

These tax exemptions mean that oil is more heavily taxed than coal and natural gas; see Fig. 13.5. The average effective tax rate for oil is nearly twice the level of coal and natural-gas taxes. This is because of the large market share for oil in transport and households, where few tax exemptions apply. Even the rate for natural gas is slightly higher than for coal. This is also the result of coal having a relatively larger market share than natural gas in sectors that enjoy tax exemptions. The tax structure chosen for the electricity sector also helps to lower the average tax rate for coal. Generally a levy on electricity sales instead of a carbon tax on fuel use is preferred, in order not to undermine the financial situation of utilities with a large share of coal-fired capacity. This also hinders increased imports of electricity from countries with lower taxes or large shares of carbon-free electricity. Consequently the tax does not provide any incentive for fuel switching in the electricity sector. Industrial energy

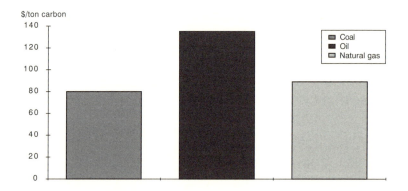

FIG. 13.5. Effective carbon tax rates by fuel, national rebound, 2020, Total Europe

consumers are not the only ones to reject high carbon taxes: perceived and actual effects on income distribution raises major public resistance to high taxes and prevents them from taking on a dominant role in climate policy.

Carbon tax levels vary between countries, depending on energy-sector structures and political factors. On average the nominal carbon tax rate in Western Europe is 150 $/ton carbon in 2020. The nominal tax is the rate initially defined by political authorities. This rate generally applies to energy consumption in households, most service sectors, and transport. The effective tax rate, i.e. the average rate after adjusting for tax exemptions, is 90 $/ton carbon, i.e. 60 per cent of the nominal rate. The tax level is considerably lower in Central Europe, whereas the effects on CO_2 emissions are larger than in Western Europe; see Fig. 13.6. CO_2 emissions decline by 20 per cent in Central Europe and by 12 per cent on average in Western Europe. The tax is less effective in Western Europe because large parts of energy consumption have been reduced through previous price and tax increases and the carbon tax: demand is relatively insensitive to further increases. Moreover, energy consumption that does not have high energy taxes is also what is exempted from the carbon tax.

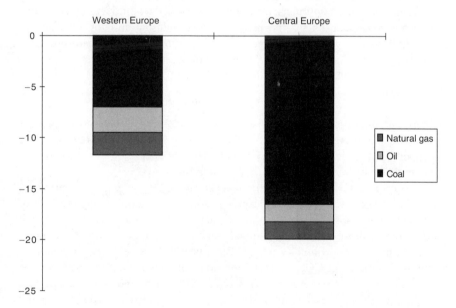

FIG. 13.6. Effects of carbon taxes on CO_2 emissions in Western and Central Europe, national rebound 2020 (% reduction)

The major part of the emission reduction in Central Europe is attributable to lower coal consumption (Figure 13.6) due to the dominant position of coal. Reduced coal consumption is also the principal source of emission reduction in Western Europe, but oil and gas still make significant contributions. Natural gas contributes as much as a quarter of the emission reduction. This may seem surprising in light of the fuel substitution towards gas triggered by the tax. However, as noted above, the configuration of the taxes provides rather weak incentives for fuel substitution. In addition, price levels for gas are lower than for oil, in turn leading to larger price implications of the tax on gas compared to oil.

Within Western Europe there are also large differences in carbon tax levels, with corresponding differences in emission reductions. Taxes are, for example, considerably higher in France than in Germany. This is partly explained by the differences in abatement costs between the two countries, see section 13.1 above. Therefore, more forceful measures are needed in France to reach the national target. In addition there is somewhat less opposition in France to carbon taxes than elsewhere, because that industry's energy use is less carbon-intensive. In addition a carbon tax has some political attractions in France since it penalizes the direct use of fossil fuels and spurs substitution towards electricity based on nuclear power. In Germany, on the other hand, energy consumption is carbon-intensive; a high carbon tax would have considerable negative consequences for the manufacturing industry and for such energy industries as coal mining and electricity generation. Taxation therefore has a relatively modest role in German climate policy.

Energy-conservation Policy

Energy-conservation policy consists primarily of the same measures that were in use in the 1970s and 1980s. As part of a climate policy strategy they are stepped up considerably. The design and effects of energy conservation instruments differ by sector.

In transport, targets for fuel efficiency in motor vehicles are the principal measures. The effect of such measures is somewhat hampered by the lack of international co-ordination. The global nature of the car industry makes it difficult to impose mandatory national standards. Efficiency targets are therefore primarily of a voluntary nature. But since all European countries pursue ambitious targets for CO_2 and NO_x emissions, there is growing pressure on the industry to produce motor vehicles that pollute less and have higher fuel efficiency. Energy-conservation considerations are also incorporated in the planning of urban development and transport infrastructure, with large programmes for public transport development. This results in major improvements in urban air quality, and at the same time makes a fairly large contribution towards reducing transport fuel use and CO_2 emissions. Such measures are particularly important in France and south-western Europe because a large part of CO_2 emissions in these countries originate from the transport sector.

Building codes are tightened up, with notable effects on energy demand in residential and service sectors. Technical standards are also imposed that improve the efficiency of heating boilers in large buildings. Financial support programmes for investments in energy efficiency are stepped up radically, for a large part financed directly by the carbon tax. Such measures have proven effective in the past and represent an important part of a nationally based energy-conservation policy. Linked to this are measures that promote efficiency improvements in electricity end-use. Demand-side management (DSM) schemes are actively pursued by energy utilities in some countries, primarily those where the carbon intensity of power production is large. DSM has only a negligible role in the liberalized electricity markets of Norway, Sweden, Finland, and the United Kingdom.

For industry the emphasis is on efficiency standards. Mandatory standards are introduced to regulate the fuel efficiency of machinery and boilers. Energy auditing of processes is also effective. However, the approach taken on this front is generally cautious, so as not to distort the competitive position of industry. Standards are therefore largely formulated as BAT (Best Available Technology) requirements, without detailed technical specifications and requirements. Voluntary agreements with industry associations on targets for energy-efficiency improvements often have a notable impact, though the announced effects may overstate the achievements that can be directly credited to the agreements.

Energy-conservation measures give a reduction of 9 per cent in CO_2 emissions for total Europe—considerably less than the carbon taxes. In Central Europe the contribution from energy conservation is only 7 per cent, despite the large potential for energy-efficiency improvements. However, the carbon taxes do spur energy efficiency, thereby leaving less to be achieved by non-pricing measures.

Fuel Substitution

The most effective fuel substitution measures are directed towards investment and disinvestment decisions in the electricity sector, though policies to restrict coal use in industry also have a clear impact on emissions in some countries. Fuel substitution measures hold a large potential for emission reductions. We may distinguish three broad categories of measures:

- Restricting coal use in power generation and district heating by prohibiting the construction of new plants, and introducing regulations that accelerate the retirement of old plants
- Promoting renewable energy sources for power and heat generation
- Constructing new nuclear capacity and extending the lifetime of existing nuclear plants

If these options were exploited in full, and electricity generation became carbon-free, emissions in 2020 would be reduced by 30 per cent. This still falls short of the target of a 34 per cent reduction. However, with the carbon tax and conservation measures specified above, a carbon-free electricity sector would yield emission reductions in excess of the target.

If the coal phase-out and the potential for renewable energy sources can be fully exploited, nuclear need not be brought into the climate policy. In most countries, nuclear is the least desirable alternative in the case of high environmental concern. For various reasons, however, neither coal-use restrictions or renewables are exploited in full (Fig. 13.7), with the implication that the nuclear option is being utilized. Moreover, some countries need not pursue

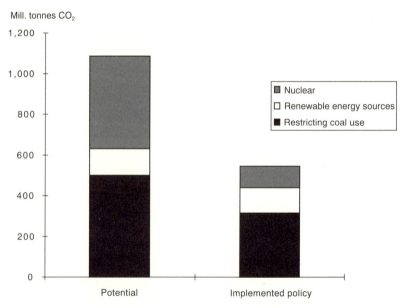

Fig. 13.7. Potential and implemented CO_2 emission reduction by fuel substitution in 2020, national rebound

coal phase-out or renewable options in full to meet their national emission targets, whereas other countries will not meet their target even after a full coal phase-out and maximum use of renewables.

Restricting Coal Use

Some two-thirds of the potential for reductions in emissions by coal phase-out is taken up. In several countries there is a complete phase-out of coal consumption. The major coal-consumers, Germany and the Visegrad countries, also see dramatic reductions in coal use, but part of coal-fired electricity generation is retained, for political and economic reasons. In Germany some coal plants based on domestic lignite and imported hard coal are kept because they are highly profitable and cause little air pollution. Security of supplies is an additional argument for not retiring the plants. If coal were phased out entirely in Germany and replaced by imported natural gas, gas imports would increase significantly; most of this import would come from Russia.

Energy security is an equally important argument in Central Europe for keeping coal plants. In the case of a coal phase-out, natural-gas imports would increase by more than 60 per cent, to a level three times higher than in 1995. In addition there are economic and social arguments for letting oil and natural gas contribute to some of the emission reductions in these countries.

Promoting Renewables

Of the technical potential for renewable energy sources (solar, wind, biomass, and hydro) some 75 per cent is exploited: 300 TWh/year of additional electricity and heat are generated in 2020. This indicates a major effort across Europe to accelerate the introduction of renewable energy sources; it comes on top of a rapid increase in the supply of renewable energy sources already seen in the case of moderate environmental concern. Still, some 100 TWh/year of the potential is left untapped, for various reasons. First, renewable energy sources that require large financial resources and radical support programmes are in some countries considered too costly compared to other alternatives. This is an important factor in Central Europe and some South European countries. Second, in Norway and Sweden the potential for additional hydro power production through lifting of river protection is only partly exploited. These electricity systems are practically carbon-free at the outset, and fuel substitution possibilities that can yield emission reductions are limited. Third, there are environmental problems associated with renewables—such as loss of biodiversity, arable land, and recreation values from hydropower construction; noise, visual impact, and safety issues from wind-power plants. This is of some importance to several Western European countries.

Retreating to Nuclear

Compared to its potential, nuclear is put to only modest use as a climate policy measure. Still, nuclear contributes as much as renewable energy sources. Nuclear is in most countries considered the 'last resort' in climate policy, brought in only after high carbon taxes have been imposed and radical programmes implemented for renewables and coal phase-out. On the other hand, in some Western European countries the nuclear option is needed. Depending on national conditions, nuclear may also be an attractive option, as it displaces CO_2 at a cost below the marginal cost of any alternative. This, together with the concern over increased reliance on natural-gas imports following restrictions on coal use, makes some Western European governments select nuclear as one alternative for curbing CO_2 emissions.

Some 20 TWh/year in additional nuclear power generation is accounted for by France. This may be considered surprisingly low, given the relatively weak public opposition to nuclear, as well as the high abatement costs in France. That nuclear makes such a moderate contribution is mostly explained by technical factors. There is very little scope for emission reductions through forcing out fossil fuels in the electricity sector, since nuclear is already dominant. A further increase in the share of nuclear in combination with growth in renewables runs counter to strong technical arguments. The growth in medium- and peak-load demand requires that the sector has a sizeable amount of capacity more flexible than nuclear and renewables. Thus there is a need to keep gas- and oil-fired generating capacity at a certain level. A rise in nuclear generation therefore requires increased electricity demand through a further shift from direct use of coal, oil, and natural gas. This shift is induced by the high carbon taxes and by regulations on coal and oil use in industry. Moreover, the close relationship between the gas and electricity monopolies of France (GdF and EdF) and the political authorities ensures smooth implementation for a strategy to substitute natural gas for electricity.

Relative Contributions from Climate Policies

Of the three main categories of climate policy measures, carbon taxes and fuel substitution measures make the largest contribution; see Fig. 13.8. In Western Europe fuel substitution is slightly more important than carbon taxes. Energy-conservation measures give a lower contribution. This is mainly caused by the difficulties of finding effective conservation measures that can be implemented within a national context. In Central Europe, low energy prices make the carbon tax particularly effective. Taxation is the principal climate policy

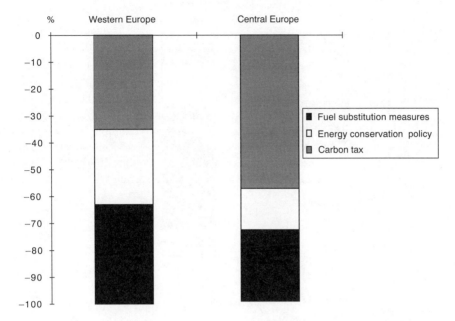

F<small>IG</small>. 13.8. Relative impact of climate policy measures by category, national rebound scenario

measure, contributing more than half of the needed reduction in these countries. Energy conservation is less important, due to the major efficiency improvements achieved by the taxes. Fuel substitution is relatively important because of a notable growth in renewables, which largely replace old and inefficient coal plants.

Impacts on NO_x and SO_2 Emissions

Climate policy measures ensure by a clear margin that SO_2 targets are met for most countries. Hence there is no need for a further reduction by mandating emission control measures to reduce SO_2 emissions, except for a few locations with high local concentration of sulphur. The fuel substitution measures alone are sufficient to fulfil the SO_2 target due to the major reductions in coal use in industry and the power sector.

For NO_x, climate policy measures are not sufficient to enable the target to be reached. Although energy-conservation policies are relatively effective in

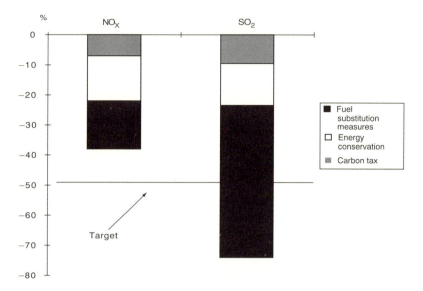

F𝙸𝙶. 13.9. Reduction in NO$_x$ and SO$_2$ emissions from climate policy measures in 2020, national rebound scenario

curbing NO$_x$ emissions, due to the emphasis given to transport sector measures, additional emission control policies have to be installed.

13.3. Climate Policy in Liberalization and Trade

A Common European Climate Policy

Policies to reach the common EU target are formulated and implemented mostly at the EU level. This is in line with the framework for policy of this scenario, where energy and environmental policy objectives are to be dealt with at the EU level rather than unilaterally by individual national governments. In view of the radical nature of the emission targets, especially for CO$_2$, the EU bodies decide on far-reaching measures that change trends and patterns in energy consumption. The EU is authorized to impose a carbon tax with large revenues accruing to the EU. Considerable funds are allocated to climate-related research, and financial support is provided, at the discretion of the EU Commission, to energy conservation and fuel substitution efforts.

The most important means of spurring energy conservation and fuel substitution, however, are through the Commission's implementation of regulations that mandate technology and fuel choices. Such regulations have a substantial influence on investments in energy-using machinery and equipment and on the competitive position of different fuels. The new and stringent regulations represent a clear break with the role of EU energy policy from the moderate concern scenario. The aim of energy policy then was to ensure economic efficiency and security in energy supplier, whereas this scenario emphasizes both supply and demand-side issues through measures that are quite forceful and even interventionist. This shift is seen as necessary in order to tackle the difficult issues of sharing the burden of emission reductions among countries. Moreover, there is a common understanding among EU member countries that the Commission needs tools to intervene in markets in order to ensure that the path towards the emission target is actually followed.

A Harmonized Carbon Tax

The general strengthening of EU bodies and the transfer of energy and environmental policies from national authorities to the EU level gives the EU Council the formal authority to impose a common carbon tax on all member countries. Several factors suggest that the carbon tax should be the principal measure in EU climate policy. First, a tax raises funds that can finance EU programmes, including research and financial support programmes for a climate policy strategy. Second, a harmonized tax does not distort trade and competition in a manner that is in conflict with the EU rules and regulations. Third, a carbon tax has attractions as a cost-efficient measure to reduce emissions.

But a carbon tax also meets with considerable political resistance. Industry and other energy consumers faced with escalating energy costs vigorously oppose a high tax. Objections are particularly strong from countries that are affected disproportionally, which is the case for Central European countries with a high coal share in energy production and consumption. Central Europe, and some countries in Western Europe, are also committed to increasing energy price and taxes through the ongoing price harmonization in liberalization and trade. Hence a carbon tax adds to the political difficulties of carrying through these price adjustments.

Countries with less carbon-intensive energy-demand patterns and/or with energy prices above the EU average are less opposed to and less affected by a high carbon tax. Countries that are exporters of low-carbon or carbon-free energy (e.g. France, Norway, and Netherlands) would partly benefit from a carbon tax through a 'wind fall profit' on their exports.

The final configuration and exact level of the harmonized tax are therefore the result of lengthy negotiations where burden-sharing in achieving the CO_2 target is the central issue. The result of this process is a carbon-tax level which gives the tax a central role in EU climate policy, but with modifications compared to a 'pure carbon tax':

1. The nominal rate is set at 230 US$/ton of carbon, which is 2.5 times the tax level proposed by the EU in 1991 for implementation in the year 2000. This means major increases in energy prices for those sectors that are not granted exemptions or lower tax rates.

2. Tax exemptions are granted to certain industries that are exposed to international competition. However, harmonization of taxes across Europe implies that the scale of exemptions is smaller than would have been the case with unilateral taxation policies. Some 30 per cent of emissions are exempted from the tax.

3. A levy is introduced on electricity sales rather than a carbon tax on fuel use in the electricity sector, so as not to alter the competitive position between power companies. Incentives to affect fuel choices for new capacity and the rate of plant retirement are pursued through regulatory measures and financial support schemes (see the discussion below).

The tax has a major impact on emission reduction in Central Europe, see Fig. 13.10. This is the result of the dominance of coal in the systems of Central Europe. In addition, energy-price levels are initially low in these countries, before the harmonization of price levels in Europe have been completed.

In Western Europe, emission reductions from the tax are considerably lower because coal holds a low share in energy consumption, and because energy-price levels are high. Still, the carbon tax makes an important contribution towards the emission target for Europe as a whole. The carbon tax contributes an emission reduction for total Europe which is more than half of the reduction required to reach the target.

Vast revenues accrue from the tax. By 2020 the tax is yielding $37 bn. Prior to 2020, before taxes and other climate measures have had full effect, annual proceeds are even greater. Although some of the revenues are recycled to member countries, a substantial part is earmarked for EU programmes that promote energy conservation and fuel substitution.

Energy-conservation Policies

Energy-conservation measures are primarily implemented through a considerable strengthening of the SAVE programme. Emphasis is on improvement in

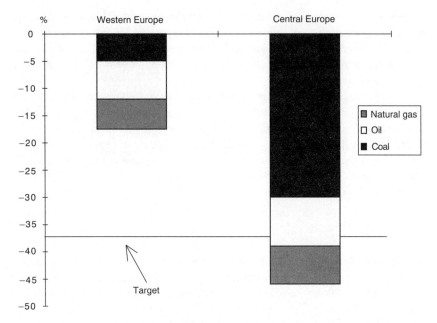

FIG. 13.10. Reductions in CO_2 emissions from carbon tax in Western and Central Europe, 2020

the energy efficiency of motor vehicles, various equipment, and electrical appliances. Results are achieved partly through mandatory standards and partly by voluntary targets agreed with industry associations. Information, 'awareness campaigns' and eco-labelling are also pursued through coordinated measures. All the regulatory steps apply equally to all countries, but there are various financial-support programmes that are allocated in accordance with an agreed burden-sharing scheme. This includes support for modernization of industries and refurbishment of heating systems for buildings. Major pro- grammes are set up for research, development, and demonstration in new energy-efficient technologies. Eventually the latter have large impacts on efficiency improvements in transportation and industry.

Much emphasis is on measures to lower transport fuel demand; the impact of these measures is significant. The rationale for emphasizing transport fuel demand is twofold. First, transport fuel consumption is a large and growing source of CO_2 emissions and the carbon tax is largely ineffective in curbing the growth, due to high tax levels at the outset. Second, the transport sector is also a significant source of NO_x emissions and is considered an important target in a strategy to reduce such emission.

Conservation in electricity demand is relatively modest because demand-side management programmes (DSM) by utilities are of minor importance. Electricity enterprises, which compete for market shares in both wholesale and retail markets, do not perceive any commercial benefits from allocating funds to DSM activities.

Energy-conservation policies account for some 6 per cent of emission reductions by the year 2020, calculated as the additional effect to the carbon tax. This may be considered modest in light of the positive impacts of coordinating conservation measures across European countries. It should however be kept in mind that the carbon tax and energy-conservation policies in combination result in a 22 per cent reduction in energy consumption, and a decline in CO_2 emissions of 27 per cent.

Fuel Substitution

Policies for fuel substitution consist of financial incentives and regulations mandating fuel and technology choices. Financial incentives are important for fuel substitution in the transport sector. On the other hand, fuel substitution in the transport sector has primarily local benefits and only contributes to small reductions in CO_2 emissions through ethanol and other fuels based on biomass. The major contribution to CO_2 emissions reduction comes from measures that induce fuel switching in electricity and district heating, and to a lesser extent in industry. The emphasis is on regulatory measures, though financial support programmes are relevant for some countries.

Four categories of measures are important:

- Directives prohibiting the construction of coal-fired power plants
- Financial incentive schemes to advance the introduction of renewable energy sources
- Stringent technical standards that accelerate the retirement of coal plants
- Financial incentives and moderate relaxation of technical standards, making it more attractive to refurbish and extend the life of nuclear plants

The most important measures are those that influence decisions to invest in new capacity. Policies towards existing plants, in order to accelerate plan retirement, are more cautious because such measures would affect countries and electricity enterprises differently. Some national governments argue against a radical approach towards existing coal plants, since this would alter the competitive position of enterprises within the electricity-supply industry. This is deemed important because there is extensive competition and trade with electricity across national borders. For the same reason it is contended that financial incentives and regulatory changes meant to delay the retirement of nuclear plants should not be tailored in a way which gives nuclear power generators a 'wind-fall profit'.

These arguments have an affect on the nature of regulations and interventions made by the EU Commission, but do not prevent fuel substitution measures from being implemented rather selectively, with important implications for competition and trade. For example, some utilities in Central Europe are given permission to refurbish and even to build new coal plants, whereas other plants in selected areas of Western Europe are faced with more stringent technical standards than the general norm, which forces plants out of operation.

In total, fuel substitution holds a potential for emission reductions far beyond the emission target, once the effects of the carbon tax and the energy-conservation policy have been taken into account; see Fig. 13.11. As mentioned, carbon taxes and energy-conservation policies in combination reduce CO_2 emissions by 27 per cent. With the reduction target set at 37 per cent some 10 per cent is

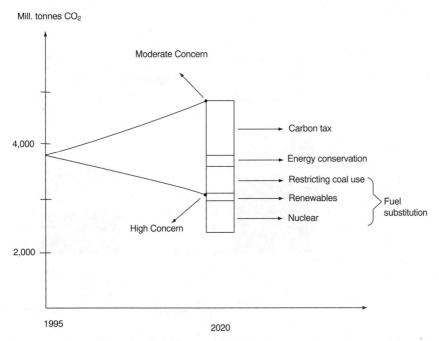

FIG. 13.11. Potential reduction in CO_2 emissions from fuel substitution in 2020, liberalization and trade

required from fuel substitution measures. Coal phase-out and renewable energy sources have the potential to achieve this (Fig. 13.11). However, a total coal phase-out is not pursued, for various political and economic reasons, and all the three categories of fuel substitution measures make a contribution. The outcome can be summarized as follows (see Fig. 13.12):

- Restrictions on coal use are being exploited only to 50 per cent of the potential.
- The potential for renewable energy sources is exploited in full.
- Nuclear attains an important role in climate policy strategy.

In order to understand these results we need to review them in the context of the entire set of climate policy measures imposed by the EU, seeing how the policies affect different countries. In this regard, the burden-sharing issue is of particular importance.

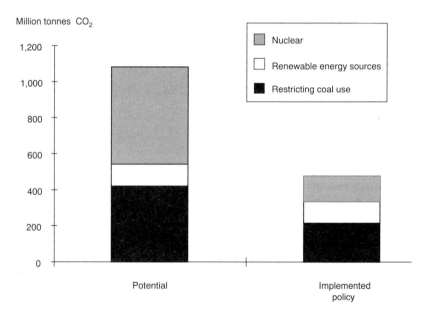

FIG. 13.12. Potential and implemented CO_2 emission reduction by fuel substitution, liberalization and trade, 2020

Note: The emission reductions contributed by each of the three options are calculated by assuming that: (i) coal is replaced by gas (CCGT) (ii) renewables replace gas (conventional and CCGT) (iii) nuclear replaces gas (CCGT).

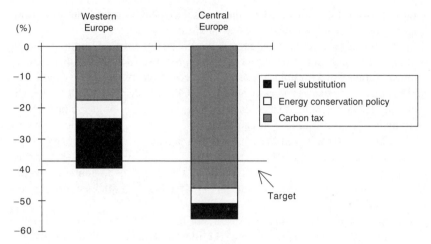

Fig. 13.13. Reduction in CO_2 emissions by policy category in 2020, liberalization and trade scenario

Burden-Sharing and the Overall Policy Mix

As noted above, the harmonized carbon tax has the greatest effect in Central European countries, with emissions being reduced by 50 per cent. By contrast, the relative reduction in Western Europe is only half of this; see Fig. 13.13. This implies that Central European and some other countries make disproportionally large contributions to the total European emission reduction. These countries are offered various forms of compensation in the form of financial transfers, for energy-sector purposes and other ends, and through exemptions from regulations and standards. Those countries who contribute modestly through the carbon tax and conservation measures are to some extent subject to interventions from the EU Commission. Such actions primarily affect the fuel mix.

Therefore carbon taxes and energy-efficiency regulations apply equally to all EU member-countries whereas regulations of fuel choices and financial-support programmes are applied selectively by the Commission. Burden-sharing has in this respect a major bearing on the implementation of measures.

Restricting Coal Use

As a result of this policy, energy companies in Central European are to some extent exempted from the regulations on coal use, and financial support is provided to modernize power and district heating plants. Coal consumption is not entirely phased out. There is, however, a reduction of nearly 70 per cent in coal use, primarily as a result of the carbon tax but also because of energy-conservation measures and EU restrictions on coal use.

As noted, there is some reluctance at the EU policy level to impose stringent technical standards on existing coal plants because this would discriminate against companies with a large portfolio of coal-fired plants. This would have serious financial consequences for the large German electricity enterprises, for example. However, some discriminatory market intervention is the only means by which the EU Commission can actually carry through its programme for emissions reduction. Given the conflicts and political stalemate on any proposal to impose even more radical technical standards or taxation that applies equally for all countries, the climate policy strategy needs to contain selective measures that can force some countries to go further than others in phasing out coal consumption.

Energy security is an additional argument for retaining some coal for power generation. As described in Chapter 11 liberalization and trade results in a major increase in imports of natural gas for electricity generation, partly because of the decline in nuclear power production and because gas is preferred over coal in new plants. This development is accentuated by the restrictions on coal use. If natural gas were to replace coal in electricity generation, the result would be a 30 per cent increase in European gas consumption for power generation, even after the carbon tax and energy-conservation programmes have lowered electricity demand. Furthermore, natural-gas consumption in the power sector would in 2020 be eight times higher than in 1995. The EU is therefore seeking means that can curb the growth in the demand for natural gas. That includes 'saving' some coal-fired plants, as well as renewables and nuclear power.

Renewables

Renewable energy sources are exploited in full. The details of a programme are worked out by the Commission, including the contribution in terms of emission reductions provided by each country and the extent of EU financing. A large part of the investment in renewables in Central Europe is financed by the EU, due to the large contribution made through carbon taxation. Programmes for renewables in Western Europe are financed mainly by national funds.

The strong emphasis on renewable energy sources means that protection orders on rivers are lifted in the Norway and Sweden. This involves some serious conflicts between climate policy objectives and other environmental concerns. However hydropower and nuclear in the case of Sweden are the only means by which these countries can make significant contributions to fuel substitution.

Nuclear Power

Throughout Europe there is massive opposition to applying nuclear as an important part of climate policy strategy. This opposition makes it especially hard to obtain approval from a majority of EU member-states to increase nuclear production. However, nuclear is brought in as a residual, bridging the gap between feasible achievements by other measures and the emission target. The EU retreats to nuclear, for the following reasons:

1. A further increase in carbon taxes is considered unacceptable, in view of the major costs involved, especially for those countries already taking a large share of the emission reduction.

2. More stringent efficiency standards are opposed, due to high costs and conflicts in burden-sharing. Increased financial-support programmes for energy efficiency are even more costly and with uncertain effects.

3. Further restrictions on coal use are not carried through, for the reasons stated above.

4. Nuclear represents a relatively attractive option in economic terms in this scenario. Moderate environmental concern in the liberalization and trade scenario indicates large retirements of nuclear plants that can be halted. The costs of upgrading and extending the lifetime of these plants are low compared to other carbon-free alternatives or energy-conservation measures over and above what has already been decided.

5. France makes a large contribution to emission reductions from nuclear. France is obliged to make additional emission reductions since the carbon tax and other EU regulations have little effect. Refurbishing nuclear plants originally due for retirement is the only means by which France can make such a contribution.

The extended use of nuclear implies that nuclear production increases by 125 TWh/year in 2020. Production after this is still some 15 per cent below the level in 1995. Hence nuclear is primarily brought in through extending the lifetime of existing plant, which is less politically controversial than finding sites and constructing new plants.

Impacts on NO_x and SO_2 Emissions

Climate policy also has major impacts on SO_2 and NO_x emissions. SO_2 emissions are reduced by 60 per cent by the year 2020, compared to moderate environmental concern, comfortably in excess of the target. The reduction is primarily a result of the radical reduction in coal consumption in the power sector and industry. These two sectors are the main sources of SO_2 emissions,

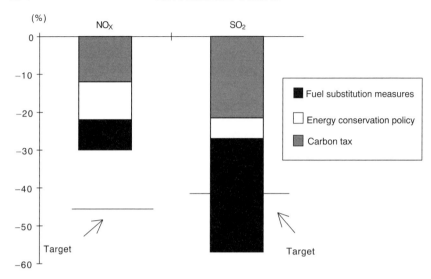

Fɪɢ. 13.14. Reduction in NO_x and SO_2 emissions from climate policy measures in 2020, liberalization and trade

and climate policies contribute to a 70 per cent reduction in coal use from the sectors.

Reductions in NO_x emissions are more moderate. Both the carbon tax and fuel substitution have less impact on NO_x emissions than on SO_2 emissions. These measures spur fuel switch from coal to natural gas, yielding major reductions in SO_2 emissions (since natural gas does not emit SO_2), whereas NO_x emissions are affected only to a limited extent. Energy-conservation policy, however, has a relatively important impact on NO_x emissions, because conservation measures are targeted largely towards transport, and are motivated partly by the NO_x target.

Our analysis has shown that climate policy measures result in SO_2 emissions reduction that exceeds the target, whereas reductions for NO_x fall ten percentage points short of the target. Additional emission control measures are therefore imposed to meet the NO_x target.

13.4. Energy Market Impact

Emission control measures for the purpose of the NO_x target have little impact on energy market developments. The large effects are from climate policy

Table 13.1. *Energy consumption in Europe in two energy scenarios and environmental futures*

	1995	Moderate Concern		High Concern	
		2020NR	2020LT	2020NR	2020LT
Coal	351	392	374	95	110
Oil	654	758	792	637	648
Natural gas	303	446	531	424	387
Nuclear power	215	273	145	300	177
Renewables	40	61	61	75	80
Total primary energy supplies	1,563	1,930	1,902	1,531	1,402
Electricity output (TWh)	2,320	3,193	3,201	2,826	2,675
CO_2 emissions (mill. tonnes)	3,769	4,553	4,788	3,008	3,008

NR=national rebound, LT=liberalization and trade

measures in the form of lower energy demand and a shift in the fuel mix of the five primary energy sources: coal, oil, natural gas, nuclear, and renewables. Total energy consumption paths of the two scenarios, with moderate and with high concern, show only small differences. The consumption level is however somewhat lower in the liberalization and trade. This implies that energy consumption is more carbon intensive in this scenario, since both scenarios end up having the same level of CO_2 emissions in 2020.

Climate policies induce major changes to the fuel mix. Coal consumption is in both scenarios reduced by more than 70 per cent in 2020, thus resulting in very low market shares; 8 and 6 per cent in liberalization and trade and national rebound respectively. The higher coal share in liberalization and trade mirrors the higher carbon intensity and is explained by the more restrictive policy on coal use.

For natural gas there are some more distinct developments between the two scenarios. In liberalization and trade natural-gas consumption declines by 25 per cent. In national rebound the reduction is only 5 per cent. There are three main factors behind the sharp decline in natural-gas consumption in liberalization and trade, all related to policies affecting fuel use in the electricity sector. First, and as noted above, restrictions on coal use are less severe in liberalization and trade than in national rebound. Therefore the substitution towards natural gas is lower. Second, carbon taxes and conservation measures have a larger effect in liberalization and trade, resulting in a considerable slowdown in the requirement for new generating capacity. Since CCGT (combined-cycle gas turbines) is the preferred technology for investment in

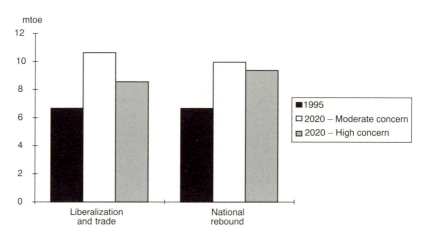

FIG. 13.15. Natural-gas consumption in Europe in the two scenarios, 1995 and 2020

the electricity sector, natural gas is particularly affected by lower electricity demand. Third, renewable energy options are pursued more aggressively, thus further reducing the investment in new gas-fired plants.

Natural gas in liberalization and trade also loses market shares in the industry and residential/service sectors, though to a lesser extent than in the electricity sector. This is partly because there is less substitution towards natural gas in

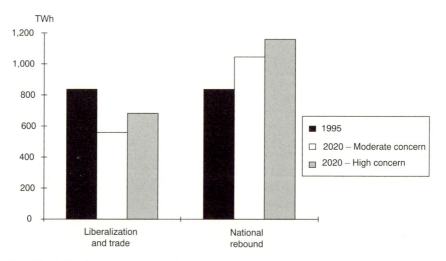

FIG. 13.16. Nuclear power generation in Europe in the two scenarios, 1995 and 2020

industry in the liberalization and trade scenario, i.e. EU measures to restrict coal use in this scenario are less stringent than national measures in the national rebound scenario. In addition the decline in energy consumption in residential/service sectors is larger in liberalization and trade, with reductions in natural gas use as a result. National rebound becomes the 'gas scenario' whereas it was the opposite in the case of moderate concern.

Oil consumption follows similar trends in the two scenarios. Consumption is reduced by 16 and 18 per cent in national rebound and liberalization and trade respectively. It is higher in liberalization and trade because the carbon tax and energy-conservation policies have a larger role in that scenario. Oil is only modestly affected by fuel substitution measures which are the principal category of measures in national rebound.

In both scenarios nuclear power production increases as a result of higher environmental concern. The additional production is higher in liberalization and trade (125 TWh/year in 2020) than in national rebound (105 TWh/year in 2020). However, national rebound continues to show much higher nuclear production than liberalization and trade.

13.5. Conclusion

High environmental concern calls for a radical shift in energy policy, with climate-change mitigation as the overriding objective. The measures that were effective in reaching the policy objectives of the 1970s and 1980s are inadequate and partly inapplicable. A variety of new instruments are required and need to be adjusted to the corporate structure and framework for policy. The scenarios presented in the preceding chapters represent two distinct energy structures and conditions for action.

Although energy systems and the corporations and interests connected with them do not alter rapidly with political change, liberalization and trade eventually heads to a major break with past structures. Policies at the European level are formulated with consideration given to potential distortive effects on competition and trade. Such policies are not necessarily fine-tuned to local interests and structural differences between countries. The old social contracts established at national levels are broken and energy enterprises are less prone to be vehicles for political ends. Policies depend more on changing the conditions for fuel and technology choices made by commercial decision-makers and consumers, particularly through taxation, subsidies, and regulations. An important feature of radical environmental policy in liberalization and trade, however, is the strong influence of burden-sharing on the design of policy measures. As the discussion above has shown, these issues can only be tackled through forceful and even interventionist measures. Hence the ambitious CO_2 reduction target conflicts with 'liberalism' and 'liberalism', to some extent, will have to yield.

Strategies developed within national rebound are, by contrast, modulated and developed on the basis of the national characteristics and the historical preconditions outlined in Part I of this book. National Rebound is in its features dirigible and relies on existing monopolies and sector interests in its approach to the new environmental challenges. Electricity enterprises are, as they were in the past, the principal instruments for change. But their contribution cannot ensure emission targets will be reached. Therefore, demand-side measures (e.g. carbon taxes) are called for. This does not subvert the structure of national rebound but it harms the market position of energy-supply enterprises.

ANNEX A

Table A.1. *Energy consumption, Europe, national rebound, moderate concern*

	Levels (mtoe)				Growth rates (% p.a.)			
	1970	1995	2005	2020	1970–1995	1995–2020	1995–2005	2005–2020
Coal	447	351	356	392	−0.9	0.4	0.1	0.7
Oil	627	654	704	758	0.2	0.6	0.7	0.5
Natural gas	73	303	393	451	5.9	1.6	2.6	0.9
Nuclear power	12	215	231	273	12.4	1.0	0.7	1.1
Renewables	28	40	43	61	1.4	1.7	0.8	2.4
Total primary energy supplies	1,186	1,563	1,726	1,930	1.1	0.8	1.0	0.7
Electricity output (TWh)	1,100	2,320	2,593	3,193	3.0	1.3	1.1	1.4
CO_2 emissions (mill. tonnes)	3,613	3,769	4,139	4,553	0.2	0.8	0.9	0.6

Table A.2. *Energy consumption, Western Europe, national rebound, moderate concern*

	Levels (mtoe)				Growth rates (% p.a.)			
	1970	1995	2005	2020	1970–1995	1995–2020	1995–2005	2005–2020
Coal	327	250	263	286	−1.0	0.5	0.4	0.6
Oil	599	620	661	697	0.1	0.5	0.7	0.4
Natural gas	63	281	364	403	6.2	1.5	2.6	0.8
Nuclear power	12	205	220	262	12.2	1.0	0.7	1.2
Renewables	27	39	42	60	1.5	1.7	0.8	2.4
Total primary energy supplies	1,028	1,395	1,550	1,708	1.2	0.8	1.1	0.6
Electricity output (TWh)	996	2,150	2,408	2,935	3.1	1.3	1.1	1.3
CO_2 emissions (mill. tonnes)	3,063	3,246	3,606	3,885	0.2	0.7	1.1	0.5

Table A.3. *Energy consumption, Central Europe, national rebound, moderate concern*

	Levels (mtoe)				Growth rates (% p.a.)			
	1970	1995	2005	2020	1970–1995	1995–2020	1995–2005	2005–2020
Coal	120	101	93	106	−0.7	0.2	−0.8	0.9
Oil	27	34	43	61	0.9	2.3	2.3	2.3
Natural gas	10	22	29	43	3.3	2.7	2.7	2.7
Nuclear power	0	10	11	11		0.3	0.6	0.0
Renewables	0	1	1	1	2.1	1.7	0.8	2.3
Total primary energy supplies	158	168	176	222	0.2	1.1	0.5	1.6
Electricity output (TWh)	104	170	184	258	2.0	1.7	0.8	2.3
CO_2 emissions (mill. tonnes)	550	523	533	668	−0.2	1.0	0.2	1.5

Table A.4. *Energy consumption, Germany, national rebound, moderate concern*

	Levels (mtoe)				Growth rates (% p.a.)			
	1970	1995	2005	2020	1970–1995	1995–2020	1995–2005	2005–2020
Coal	148	98	95	106	−1.6	0.3	−0.3	0.7
Oil	138	136	144	147	−0.1	0.3	0.6	0.1
Natural gas	12	67	82	91	7.0	1.3	2.0	0.8
Nuclear power	2	40	41	44	13.5	0.4	0.2	0.5
Renewables	2	2	2	2	0.1	1.7	0.8	2.3
Total primary energy supplies	301	342	364	391	0.5	0.5	0.6	0.5
Electricity output (TWh)	254	474	517	632	2.5	1.2	0.9	1.4
CO_2 emissions (mill. tonnes)	976	884	932	1,002	−0.4	0.5	0.5	0.5

Table A.5. *Energy consumption, France, national rebound, moderate concern*

	Levels (mtoe)				Growth rates (% p.a.)			
	1970	1995	2005	2020	1970–1995	1995–2020	1995–2005	2005–2020
Coal	37	19	19	19	−2.7	0.1	0.0	0.1
Oil	94	87	95	101	−0.3	0.6	0.9	0.4
Natural gas	8	29	40	45	5.2	1.7	3.2	0.8
Nuclear power	1	86	95	116	17.6	1.2	1.0	1.3
Renewables	5	5	6	8	0.1	1.9	2.4	1.6
Total primary energy supplies	145	228	255	289	1.8	1.0	1.1	0.9
Electricity output (TWh)	120	354	404	493	4.4	1.3	1.3	1.3
CO_2 emissions (mill. tonnes)	414	358	402	432	−0.6	0.8	1.2	0.5

Table A.6. *Energy consumption, Europe, liberalization and trade, moderate concern*

	Levels (mtoe)				Growth rates (% p.a.)			
	1970	1995	2005	2020	1970–1995	1995–2020	1995–2005	2005–2020
Coal	447	351	312	374	−1.0	0.3	−1.2	1.2
Oil	627	654	716	792	0.2	0.8	0.9	0.7
Natural gas	73	303	410	531	5.9	2.3	3.1	1.7
Nuclear power	12	215	219	145	12.4	−1.6	0.2	−2.7
Renewables	28	40	43	61	1.5	1.7	0.8	2.3
Total primary energy supplies	1,186	1,563	1,700	1,902	1.1	0.8	0.8	0.8
Electricity output (TWh)	1,100	2,320	2,543	3,201	3.0	1.3	0.9	1.5
CO_2 emissions (mill. tonnes)	3,613	3,769	4,052	4,788	0.2	1.0	0.7	1.1

Annex A

Table A.7. *Energy consumption, Western Europe, liberalization and trade, moderate concern*

	Levels (mtoe)				Growth rates (% p.a.)			
	1970	1995	2005	2020	1970–1995	1995–2020	1995–2005	2005–2020
Coal	327	250	229	277	−1.1	0.4	−0.9	1.3
Oil	599	620	673	726	0.1	0.6	0.8	0.5
Natural gas	63	281	380	475	6.2	2.1	3.1	1.5
Nuclear power	12	205	209	136	12.2	−1.6	0.2	−2.8
Renewables	27	39	43	60	1.5	1.7	0.8	2.3
Total primary energy supplies	1,028	1,395	1,533	1,675	1.2	0.7	0.9	0.6
Electricity output (TWh)	996	2,150	2,355	2,916	3.1	1.2	0.9	1.4
CO_2 emissions (mill. tonnes)	3,063	3,246	3,551	4,114	0.2	1.0	0.9	1.0

Table A.8. *Energy consumption, Central Europe, liberalization and trade, moderate concern*

	Levels (mtoe)				Growth rates (% p.a.)			
	1970	1995	2005	2020	1970–1995	1995–2020	1995–2005	2005–2020
Coal	120	101	83	96	−0.7	−0.2	−1.9	1.0
Oil	27	34	43	66	0.9	2.7	2.4	2.9
Natural gas	10	22	30	55	3.3	3.8	3.3	4.1
Nuclear power	0	10	10	8	—	−0.7	0.0	−1.2
Renewables	0	1	1	1	2.1	1.7	0.8	2.3
Total primary energy supplies	158	168	167	227	0.2	1.2	0.0	2.1
Electricity output (TWh)	104	170	188	285	2.0	2.1	1.0	2.8
CO_2 emissions (mill. tonnes)	550	523	501	675	−0.2	1.0	−0.4	2.0

Table A.9. *Energy consumption, Germany, liberalization and trade, moderate concern*

	Levels (mtoe)				Growth rates (% p. a.)			
	1970	1995	2005	2020	1970–1995	1995–2020	1995–2005	2005–2020
Coal	148	98	89	97	−1.6	0.0	−1.0	0.6
Oil	138	136	147	152	−0.1	0.5	0.8	0.2
Natural gas	12	67	87	106	7.0	1.9	2.6	1.4
Nuclear power	2	40	39	29	13.5	−1.3	−0.3	−1.9
Renewables	2	2	2	2	0.1	1.7	0.8	2.3
Total primary energy supplies	301	342	362	386	0.5	0.5	0.6	0.4
Electricity output (TWh)	254	474	507	624	2.5	1.1	0.7	1.4
CO_2 emissions (mill. tonnes)	976	884	927	1,018	−0.4	0.6	0.5	0.6

Table A.10. *Energy consumption, France, liberalization and trade, moderate concern*

	Levels (mtoe)				Growth rates (% p. a.)			
	1970	1995	2005	2020	1970–1995	1995–2020	1995–2005	2005–2020
Coal	37	19	18	28	−2.7	1.7	−0.2	3.0
Oil	94	87	96	107	−0.3	0.8	1.0	0.7
Natural gas	8	29	38	58	5.2	2.8	2.8	2.8
Nuclear power	1	86	95	69	17.6	−0.9	0.9	−2.1
Renewables	5	6	6	9	0.7	1.7	0.8	2.3
Total primary energy supplies	145	228	254	271	1.8	0.7	1.1	0.4
Electricity output (TWh)	120	354	394	445	4.4	0.9	1.1	0.8
CO_2 emissions (mill. tonnes)	414	358	404	518	−0.6	1.5	1.2	1.7

Table A.11. *Energy consumption, Europe, national rebound, moderate (M) and high (H) concern*

	Levels (mtoe)			Growth rates (% p.a.)	
	1995	2020M	2020H	1995–2020M	1995–2020H
Coal	351	398	95	0.4	−5.1
Oil	654	758	637	0.6	−0.1
Natural gas	303	446	424	1.6	1.4
Nuclear power	215	273	300	1.0	1.3
Renewables	40	60	75	1.7	2.6
Total primary energy supplies	1,563	1,930	1,531	0.8	−0.1
Electricity output (TWh)	2,320	3,193	2,826	1.3	0.8
CO_2 emissions (mill. tonnes)	3,769	4,553	3,008	0.8	−0.9

Table A.12. *Energy consumption, Western Europe, national rebound, moderate (M) and high (H) concern*

	Levels (mtoe)			Growth rates (% p.a.)	
	1995	2020M	2020H	1995–2020M	1995–2020H
Coal	250	286	56	0.6	−5.8
Oil	620	697	586	0.5	−0.2
Natural gas	281	403	366	1.5	1.1
Nuclear power	205	262	289	1.0	1.4
Renewables	39	60	73	1.7	2.5
Total primary energy supplies	1,395	1,708	1,369	0.8	−0.1
Electricity output (TWh)	2,150	2,935	2,628	1.3	−0.4
CO_2 emissions (mill. tonnes)	3,246	3,885	2,591	0.7	−0.9

Table A.13. *Energy consumption, Central Europe, national rebound, moderate (M) and high (H) concern*

	Levels (mtoe)			Growth rates (% p.a.)	
	1995	2020M	2020H	1995–2020M	1995–2020H
Coal	101	106	38	0.2	−3.8
Oil	34	61	51	2.3	1.6
Natural gas	22	43	59	2.7	4.0
Nuclear power	10	11	12	0.3	0.6
Renewables	1	1	3	1.7	5.4
Total primary energy supplies	168	222	162	1.1	−0.1
Electricity output (TWh)	170	258	198	1.7	−1.0
CO_2 emissions (mill. tonnes)	523	668	418	1.0	−0.9

Table A.14. *Energy consumption, Germany, national rebound, moderate (M) and high (H) concern*

	Levels (mtoe)			Growth rates (% p.a.)	
	1995	2020M	2020H	1995–2020M	1995–2020H
Coal	98	106	39	0.3	−3.6
Oil	136	147	128	0.3	−0.2
Natural gas	67	91	98	1.3	1.5
Nuclear power	40	44	46	0.4	0.6
Renewables	2	2	6	1.7	5.6
Total primary energy supplies	342	391	317	0.5	−0.3
Electricity output (TWh)	474	632	526	1.2	0.4
CO_2 emissions (mill. tonnes)	884	1002	707	0.5	−0.9

Table A.15. *Energy consumption, France, national rebound, moderate (M) and high (H) concern*

	Levels (mtoe)			Growth rates (% p.a.)	
	1995	2020M	2020H	1995–2020M	1995–2020H
Coal	19	19	5	0.1	−4.9
Oil	87	101	83	0.6	−0.2
Natural gas	29	45	28	1.7	−0.2
Nuclear power	86	116	121	1.2	1.4
Renewables	6	8	11	1.3	2.6
Total primary energy supplies	228	289	249	1.0	0.4
Electricity output (TWh)	354	493	478	1.3	1.2
CO_2 emissions (mill. tonnes)	358	432	288	0.8	−0.9

Table A.16. *Energy consumption, Europe, liberalization and trade, moderate (M) and high (H) concern*

	Levels (mtoe)			Growth rates (% p.a.)	
	1995	2020M	2020H	1995–2020M	1995–2020H
Coal	351	374	110	0.3	−4.5
Oil	654	792	648	0.8	0.0
Natural gas	303	531	387	2.3	1.0
Nuclear power	215	145	177	−1.6	−0.8
Renewables	40	61	80	1.7	2.8
Total primary energy supplies	1,563	1,902	1,402	0.8	−0.4
Electricity output (TWh)	2,320	3,201	2,675	1.3	−0.7
CO_2 emissions (mill. tonnes)	3,769	4,788	3,008	1.0	−0.9

Table A.17. *Energy consumption, Western Europe, liberalization and trade, moderate (M) and high (H) concern*

	Levels (mtoe)			Growth rates (% p.a.)	
	1995	2020M	2020H	1995–2020M	1995–2020H
Coal	250	277	82	0.4	−4.4
Oil	620	726	606	0.6	−0.1
Natural gas	281	475	350	2.1	0.9
Nuclear power	205	136	169	−1.6	−0.8
Renewables	39	60	77	1.7	2.8
Total primary energy supplies	1,395	1,675	1,285	0.7	−0.3
Electricity output (TWh)	2,150	2,916	2,494	1.2	−0.6
CO_2 emissions (mill. tonnes)	3,246	4,114	2,712	1.0	−0.7

Table A.17. *Energy consumption, Central Europe, liberalization and trade, moderate (M) and high (H) concern*

	Levels (mtoe)			Growth rates (% p.a.)	
	1995	2020M	2020H	1995–2020M	1995–2020H
Coal	101	96	28	−0.2	−5.0
Oil	34	66	42	2.7	0.8
Natural gas	22	55	36	3.8	2.1
Nuclear power	10	8	8	−0.7	−0.7
Renewables	1	1	3	1.7	5.5
Total primary energy supplies	168	227	117	1.2	−1.4
Electricity output (TWh)	170	285	180	2.1	−1.8
CO2 emissions (mill. tonnes)	523	675	296	1.0	−2.3

Table A.19. *Energy consumption, Germany, liberalization and trade, moderate (M) and high (H) concern*

	Levels (mtoe)			Growth rates (% p.a.)	
	1995	2020M	2020H	1995–2020M	1995–2020H
Coal	98	97	30	0.0	−4.6
Oil	136	152	131	0.5	−0.1
Natural gas	67	106	97	1.9	1.5
Nuclear power	40	29	37	−1.3	−0.3
Renewables	2	2	6	1.7	5.6
Total primary energy supplies	342	386	300	0.5	−0.5
Electricity output (TWh)	474	624	503	1.1	0.2
CO_2 emissions (mill. tonnes)	884	1,018	679	0.6	−1.0

Table A.20. *Energy consumption, Germany, liberalization and trade, moderate (M) and high (H) concern*

	Levels (mtoe)			Growth rates (% p.a.)	
	1995	2020M	2020H	1995–2020M	1995–2020H
Coal	19	28	14	1.7	−1.1
Oil	87	107	94	0.8	0.3
Natural gas	29	58	14	2.8	−2.9
Nuclear power	86	69	84	−0.9	−0.1
Renewables	6	9	13	1.7	3.2
Total primary energy supplies	228	271	219	0.7	−0.2
Electricity output (TWh)	354	445	396	0.9	0.5
CO_2 emissions (mill. tonnes)	358	518	320	1.5	−0.4

ANNEX B

ECON-ENERGY: Summary of Model Structure

B.1. Background

ECON-ENERGY was developed in 1990 as part of a project to analyse the impact of various kinds of greenhouse abatement policies on the international markets for oil and gas. The formal model was primarily designed to analyse the interplay between economic growth, energy markets, and CO_2 emission, stressing in particular the possibilities for substitution between different energy carriers. This annex gives a brief presentation of its structure. Strictly, ECON-ENERGY is a *demand model for the global energy market*, with energy prices determined either exogenously or by cost estimates of supplying energy to the consumers. The model system was set-up for general analysis region by region, originally developed for nine regions which added up to a global energy balance. The model can be adapted to perform analysis on a country-by-country basis by incorporating estimates of national instead of regional model parameters. For this project separate submodels have been developed for Germany, France, North-West Europe, South-West Europe, Poland, Czech and Slovak Republics, Hungary.

B.2. The Model Structure: An Overview

In developing the model, the following considerations were important.

- The model should be transparent.
- It should take the long-term nature of the problem into consideration.
- The model should include description of fuel substitution.
- It should specify relevant policy instruments for implementing various kinds of climate policy.

These requirements were kept in mind when developing the demand model. The description of technology and behaviour at various points is rather brief. Furthermore, the model structure is strongly influenced by data availability. Central parameters in the demand equations of the model, such as price- and income elasticities, do not all result from any systematic econometric efforts. Structural parameters have been assessed, based on a general evaluation of energy structures in various regions. Where available, estimates of price and income elasticities of energy demand resulting from other econometric studies are utilized in this process.

A sketch of the model structure of ECON-ENERGY (an arbitrary region) is given in Figure B.1. The model describes energy balances by distinguishing between five primary energy products: oil (OIL), gas (GAS), coal (COA), hydro (HYD), and nuclear (NUC). At the end-use level, oil consumption is further separated into four products (motor gasoline (MOO), heavy fuel oil (EWO), jet fuel (JET), and middle distillates (OPR)). The central part of the model is constituted by a set of demand relations. Transportation fuels are treated separately from the stationary energy demand. The latter does not include electricity demand, which is modelled by a single demand equation. An alternative specification would of course have been to include electricity in 'stationary (secondary) energy use' and to derive electricity use by a fuel share equation.

However, in most end uses electricity can only to a very limited degree be substituted against other energy goods; electricity demand is more typically driven by applications where electricity is the only alternative. The demand for jet fuels is modelled separately for each region, while a simple aggregate (global) relation for the development in bunker oils is specified. Since the model main application has been to analyse the future development of CO_2 emissions resulting from combustion of fossil fuels, the use of petroleum products in petrochemical production is treated separately. This submodel is also indicated in the Figure.

In the submodel for electricity production, the volumes of electricity based on hydro and nuclear power are given exogenously, as is also the net export from the region. For any given level of demand for electricity, a fuel-share model allocates the total volume of conventional thermal power between fuel oil, coal, and natural gas (for details, see below). The electricity price is determined by the unit costs of electricity generation based on fossil fuels, hydro, and nuclear—weighted by their respective shares. Thus, both supply and demand for electricity are specified in ECON-ENERGY, representing a simultaneous element in the model structure.[1]

The transformation of crude oil to products in the refinery sector is given a simple treatment; exogenously given refinery costs and distribution mark-ups are included in price relations for oil products (see indicated block in the Figure). The assessed cost structure in the refinery sector affects significantly the extent to which crude oil price or energy taxes influence end-user prices.

Taxes on energy, including a specified set of CO_2 taxes, are important policy instruments in ECON-ENERGY. Other important exogenous variables are listed in boxes to the left in Figure B.1. The assumptions made about GDP-growth are obviously decisive for the future energy picture. Equally important are the assumptions made about autonomous changes in energy technologies in different sectors.

Finally, the model calculates emissions of CO_2 from the burning of fossil fuels in

---

[1] In the operational model, this link is avoided by lagging the fuel shares in the expression for the electricity price by one year (see below).

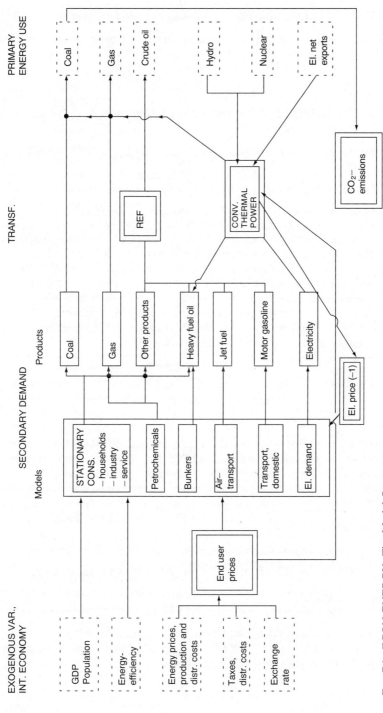

Fig. B1. ECON-ENERGY. The Model Structure

various sectors. The emission coefficients applied in these relations vary between different fuels, but the coefficients are identical for all regions.

B.3. The Submodels for the Various Demand Segments

In this section, a brief presentation of the formal structure of ECON-ENERGY is given.

B.3.1. Price Equations

End-user prices in ECON-ENERGY are calculated from estimates of costs or prices for primary energy, transformation and distribution costs, energy taxes, and exchange rates. The description of 'price formation' at various points is necessarily rather crude. The picture is severely complicated by the fact that the energy sector in most countries is subject to extensive regulation by the authorities. Prices are commonly administered by the government and may deviate significantly from marginal costs. Furthermore, price discrimination is widespread. Thus, it is not an easy task to trace the links between international prices on, say, crude oil and coal, and end-user prices in a specific region. In ECON-ENERGY, the model builder is given the option to determine on a regional basis and for each energy product whether end-user prices are driven by international markets or by domestic costs.

Import Prices

$$POILCIF = troil*TOIL + (1 - troil)*POILINT$$

$$PCOACIF = trcoa*TCOA + (1 - trcoa)* PCOAINT$$

Import prices (cif) for crude oil (POILCIF) and coal (PCOACIF) are determined by prices on the international market (POILINT, PCOAINT) and transport costs (TOIL, TCOA). Since the price variables are normalized to 1 in the base year, the weights (troil, trcoa) are interpreted and calculated as transport costs relative to the respective cif-prices.

Acquisition Costs

$$POILC = korrpo*(wo1* PPOILCIF(0)* POILCIF* V + (1 - wo1)* COIL)$$

$$PGASC = korrpg* (wg1*(PPORS(0)* POPRS - tCO2OPRSi*afraOPR/365) + wg2$$

$$*(PPHFOG(0)* PHFOG - tCO2HFOSi* alfaHFO/365) + wg3* CGAS$$

$$PCOAC = korrpc* (wc1* PPCOACIF(0)* PCOACIF* V + (1 - wc1)* CCOA)$$

For oil and coal, acquisition costs are determined as weighted averages of import prices and domestic production cost (COIL, CCOA). The latter variables cover extraction costs (per unit of energy) and transport costs to a central point in the region.

Natural gas is not traded internationally to the same extent as oil. For most regions in the model gas imports may be neglected and acquisition costs can be determined solely from the costs of supplying gas domestically. The general expression for the gas acquisition price (PGASC) is thus specified as a weighted sum of three terms; the prices of heavy fuel oil (PHFOS) and other products (POPRS) and domestic production costs (CGAS)[2]. For many of the regions in the model the weight for domestic unit costs (wg3) is set equal to 1.

End-user Prices

End-user prices for oil products are calculated as:

$$Pji = (hpji* POILC + hcji* CREFj + hdji* CDISji)*(1 + betaoil* Tji)$$

$$+ (tCO2ji*alfaj)/(365*PPji(0)), j = MOG, HFO, OPR, JET$$

Pji is the price of oil product j consumed in sector i. It is defined as the sum of crude costs, refinery costs (CREFj), a mark-up in distribution (CDISji), and taxes imposed on this delivery. Taxes consist of two elements. Existing energy taxes (Tji) are specified as value taxes, while the latter term adds emission taxes to the price. tCO_2 is the tax per unit of CO_2 released, and alfaj is the emission coefficient for oil product j. PPji is the US$-price corresponding to the end-user price index Pji. The specification of the CO_2 tax term is also affected by the fact that volumes in ECON-ENERGY are measured in million barrels oil equivalents per day (mboed), while prices are defined per barrel. As mentioned, prices are defined as indices equal to 1 in the base year, so that the 'weights' of the different costs components (hp, hc, hd) are normalized relative to the actual end-user price.

End-user prices for gas and coal are calculated by similar specifications as for oil, i.e.

$$PGASi = (hpGASi*PGASC + hdGASi*CDISGASi)*(1 + *betagas*TGASi)$$

$$+ (tCO2GASi* alfaGAS) / (365* PGASC(0) / (hpGASi*korrpg)))$$

$$PCOAi = (hpCOAi* PCOAC + hdCOAi* CDISCOAi)*(1 + betacoa* TCOAi)$$

$$+ (tCO2COAi* alfaCOA) / (365*PCOAC(0) / (hpCOAi* korrpc)))$$

B.3.2. Stationary Consumption

Stationary demand for fossil energy outside the power sector is modelled in two steps. First, total energy consumption is derived. Given a certain level of total fossil-energy use, the demand for individual fuels is determined by a fuel share model. The demand for total fossil energy per capita is represented by a log linear function:

$$TENS/POP = ast0* EFG* PFOS^{ast11}* Y^{ast2}* MSH^{ast4}* URB^{ast5}$$

$$*(TENS/POP)(-1)^{ast3}$$

---

[2] The acquisition cost price is related to oil prices <u>excluding emission taxes</u>. This is done in order to avoid double counting of CO_2 taxes in the case of natural gas.

EFG is a measure of energy efficiency. In addition, relative energy price (PEOS), the income level (GDP), the industry share of GDP (MSH), and the urbanization rate (URB) are included as explanatory variables. The price elasticity (ast 1) and the income elasticity (ast2) are interpreted as yielding short-run effects. Corresponding long-run elasticities are obtained by dividing ast1 and ast2 by the 'lag parameter' $(1 - ast3)$.

Total use of fossil fuels (TENS) includes some portion of non-commercial fuels (TRAS). The remaining ('commercial') stationary consumption of fossils is distributed among different fuels by the following relations.

$$SjS = (ssOj*PjS^{-rsj}) / \Sigma \, ssOi*PiS^{-rsi}, \; i, j = COA, GAS, HFO, DPR$$

These equations are derived from a so-called 'logit' model, where the shares may be interpreted as observations of probabilities of individual fuel choices.[3] In the logit model price elasticities are not constants, as shown by the following expressions:

$$dlog(SjS*)dlog(Pjs) = -rsj'*(1 - SjS*)$$

$$dlog(SjS*) / dlog(PiS) = rsi*SiS*$$

The (long-run) fuel elasticities (for given level of total energy use) are dependent on market share; the direct elasticity decreases with increasing share, reflecting some sort of saturation effect in the energy demand structure.

In ECON-ENERGY the fuel shares are interpreted as long-run (equilibrium) values. Actual fuel shares (SjS) in a specific period are determined by a partial adjustment mechanism, i.e.

$$SjS - SjS(-1) = lambdas*(SjS* - SjS(-1))$$

The average energy price in stationary consumption is determined by individual fuel prices weighted by fuel shares calculated by the logit model. More precisely, the average price (PFOS) is given by

$$PFOS = \Sigma \, korsi* \, SiS* \, PiS$$

where korsi is the base-year value for the relative price of fuel i (normalized against the average energy price). By this formulation, the average energy price may change as a result of changes in the fuel mix, even though individual fuel prices are constant.

B.3.3. Electricity Demand

The level of electricity use affects emissions of CO_2 indirectly via the transformation of fossil fuels in power generation. As mentioned above, we have chosen to model electricity demand independently from other stationary energy use. The demand is driven primarily by income growth, but direct price effects are also included. Again,

---

[3] It may be noted that the logit model is not necessarily consistent with traditional neoclassical assumptions. For instance price homogenity requires the following restriction on the parameters: $- rsj + rsi*Si = 0$

a constant elasticity function is used to describe electricity consumption (ELE) per capita:

$$ELE/POP = AEE0*aee0* PELE^{aee1}*(GDP/ POP)^{aee2}$$

$$*URB^{aee4}*((ELE / POP)(- 1)^{aee3}$$

Here AEE0 is an index for energy efficiency, while PELE is the relative price of electricity in the region. As mentioned in the previous section, the latter variable is determined endogenously in the model. Parallel to the end-user prices of fossil fuels, the electricity price is a weighted average of an acquisition cost (PG) and distribution costs (CDISELE), with a correction for taxes on electricity deliveries (TELE).

$$PELE= (PG*dkG1 + dkG2*CDISELE)*(1 + TELE)$$

For determining PG, we have to turn to the submodel for electricity generation.

B.3.4. Electricity Generation

The starting-point of this submodel is the link between the electricity demand just presented and the level of fossil-fuel generating capacity needed to provide this level of consumption. The total production of thermal power (ETP) is determined by the following definitional equation expressing balance of supply and demand in the electricity market:

$$ETP = (ELE + NEX)* (1 + tap) - (kapH*HYD + kapN*NUC)$$

where NEX is net exports from the region and tap is thermal losses in transmission of electricity. Finally, the production of hydro (HYD) and nuclear power (NUC) are given exogenously to determine the production of thermal power.

The next task is then to distribute the thermal power production among coal, gas, and oil respectively, i.e. to model the fuel shares. The present specification deviates from that of stationary consumption in the following way. Only new capacities, i.e. gross investments (DELTP), are assumed to be flexible and thus open for fuel choice. This means that existing capacities are assumed to be operated, but capacities decay (depreciate) over time at a rate δ.

$$DELTP(t) = ETP(t) - ETP(t - 1) + DELTP(t) = ETP(t) - ETP(t - 1) + \delta * ETP(t - 1)$$

Gross investments are then distributed among fossil fuels according to a logit model. More specifically, the shares in the new capacity (Sj) are given by

$$Sj = (sg0j* CjG^{-rj}) / \Sigma sg0i* CiG^{-ri}, i,j = COA, GAS, HFO$$

CjG is marginal costs of fuel j, which in turn is determined by the fuel price, energy efficiency (kvj), and capital costs (coj).[4]

---

[4] The term korrGj is introduced in order to correct for the fact that the fuel prices are indices; korrGj is in other words the US$ price of the fuel.

$$CjG = korrGj*PjG/kvj + coj$$

The consumption of fossil fuel x in the electricity sector (XG) is then calculated as

$$XG(t) = XG(- 1)(1 - \delta) + Sx*DELTP/kvx$$

It is probably too restrictive to assume no possibilities of substitution in the existing (surviving) thermal power capacity. One obvious aspect is the existence of dual fuel capability, which has increased its penetration in recent years. However, to take account of this in a satisfactory way would require a much more sophisticated framework. Moreover, when running the model thirty to forty years ahead, it may be argued that the total power capacity will over time be dominated by new investments.

The fact that ECON-ENERGY distinguished between new and 'old' capacities has an important implication, especially when undertaking long-term simulations. In many regions, the fuel shares in new thermal power plants differ significantly from the average fuel shares. As time passes, and old vintages of power plants are scrapped while new capacities are added to the total stock, the average fuel shares may change markedly over time. This 'demographic' effect, which can occur even though prices and costs do not change, turns out to be very important in many regions.

The acquisition cost for electricity depends on the mix of various energy carriers in the sector as a whole through the average fuel shares (SjG). As mentioned above, to avoid the problem of simultaneity, these shares are lagged one year in the price formula, i.e.

$$PG = korrG* \Sigma \ SjG(- 1)*CjG(- 1)$$

With given capacities (and thus shares) for hydro and nuclear, the average share for (fossil) fuel x (SxG) is given by

$$SxG= (1 - (SHYDG + SNUCG))*((XG*kvx)/ETP)$$

B.3.5. Transport

In a model designed to analyse environmental impact of energy consumption, it is essential to give particular emphasis to the transport sector. With continued economic growth there will necessarily be a take-off in the demand for transport fuels in developing countries. Without a breakthrough in fuel technology, it will thus be extremely difficult to avoid a dramatic increase in CO_2 emissions in these regions.

The submodels for road transport and air transport are specified for each region. An aggregate relation for ocean transport ('bunkers') is also included in the model.

Road Transport

The submodel for fuel demand in road transport starts from the following definitional relationship:

$$MOG = gamt*NCA*TVC*EFF$$

This relation separates the demand for motor gasoline (MOG) into three parts; the number of cars (NCA), the distance driven per car (TVC), and fuel efficiency (EFF). The data used to calibrate these variables all refer to private cars. MOG on the other hand, measures total energy use in the domestic transport sector. The term gamt corrects for this deviation, and for differences in units of measurement.[5]

The demand for road transport fuels is then determined by 'behavioural' equations for NCA, TVC, and EFF.

$$EFF = ATE0 * PMOG^{ett1} * EFF(- 1)^{ett3}$$

Fuel efficiency is partly price-induced (PMOG) and partly autonomous, expressed by the term ATEO. In addition, a lag effect is assumed to affect the introduction of new fuel efficiency in road transport. It should be noted that even though this equation is concave (with reasonable values on the parameters), the calculated fuel efficiency may take on unrealistic values, since there is nothing limiting its development. To avoid this problem is, however, left to the model user.

For distance driven the following relation is specified:

$$TVC = dtt0 * PMoG^{dtt1} * Y^{dtt2} * TVC(- 1)^{dtt3}$$

'Miles per car' is assumed to be dependent on the fuel price (PMOG) and the level of income per capita (Y). Again, a lag effect is introduced to distinguish between short- and long-term effects. To assume constant elasticities in the responses of distance travelled in both price and income effects may involve problems. In a low-income country the utilization of the car park is typically rather high. When income and the car park grow, the miles driven per car may actually decrease over some income interval. The same kind of effect may occur in more developed countries in a situation where a growing share of the population acquires a second car.

In the present version of the model, the path for the number of cars is determined exogenously.

Air Transport

The per capita energy demand in air transport (JET) is calculated in each region as a simple loglinear function of the price of jet fuels (PJET) and per capita income in the region, i.e.

$$JET/POP = fly0 * PJET^{fly1} * (GDP/POP)^{fly2} * ((JET/POP)(- 1))^{fly3}$$

Bunkers

To model region-specific demand equations for air transport involves the problem of capturing interregional travel in a satisfactory way. This problem is even more obvious for ocean transport; no close relationship exists between the activity level of a region and the sales figures of bunkers registered in the same region. At this point we have therefore chosen to specify a simple aggregated equation for the demand for bunker oils

[5] Note e.g. that EFF is measured in litres/km, while MOG is measured in mboed.

(HFOB). As explanatory variables we have introduced the global GDP level (GDPG) and the international crude oil price (POILINT).

$$HFOB = bbb0*(POILINT + (tCO2HFOB* alfaHFO (365* PPOILINT(0))))^{bbb1}$$

$$*GDPG^{bbb2} * HFOB(-1)^{bbb3}$$

B.3.6. Petrochemical Industry

In petrochemical production, oil products and gas are not burnt but rather used as raw materials. Consequently, no emissions of CO_2 are released. Still, as the model framework is intended to be used to analyse the implications of climate policy in the oil and gas markets, it is necessary to include the whole energy balance, including the demand for feedstock.

The energy demand in petrochemical industry is modelled in two steps. First, the total energy input (NON) in the sector is determined, as

$$NON = app0*(ORE + GRE)*(POPRP^{app1} * GDP^{app2} * NON(-1)^{app3})$$

Total feedstock is assumed to be proportional with the total reserves of oil (ORE) and gas (GRE) in the region. The actual utilization of these reserves depends on the relative energy price (POPRP) and GDP.[6]

The distribution of the total demand for feedstock is related to the composition of petroleum reserves in the region. More precisely, for given energy prices the oil (OPRP) and gas (GASP) shares are proportional to the corresponding shares of energy reserves. Changes in the prices of oil (POPRP) and gas (PGASP) will however induce changes in the fuel mix, as expressed by the following relations

$$GASP/NON = bpp0*[(POPRP/ PGASP)^{bpp1}*(GRE/ (ORE + GRE))]$$

$$OPRP = NON-GASP$$

---

[6] The price variable POPRP is the price of oil products consumed in this sector. As NON includes the consumption of both gas and oil as feedback, strictly the average energy price should have been specified in this relation. In many regions oil is the dominating input in petrochemical production, so we do not believe much is lost by this shortcut.

REFERENCES

Official Documents

COM/89/0332 final: *Draft Council Directive concerning a Community procedure to improve the transparency of gas and electricity prices charged to industrial end-users.* European Commission, 1989.

COM/89/0334 final: *Communications from the Commission. Towards completion of the internal market for natural gas. Proposal for Council directive on the transit of natural gas through the major systems.* European Commission, 1989.

COM/89/0335 final: *Draft Council Regulation (EEC) amending Regulation (EEC) No. 1056/72 on notifying the Commission of investment projects of interest to the Community in the petroleum, natural gas and electricity sectors.* European Commission, 1989.

COM/89/0336 final: *Communications from the Commission. Increased intra-Community electricity exchanges: a fundamental step towards completing the internal energy market. Proposal for a directive on the transit of electricity through transmission grids.* European Commission, 1989.

Other Works

Aarhus, Knut, and Per Ove Eikeland (1993): *Policy Measures in Pollution Abatement: Sweden, Germany, the Netherlands, Great Britain and the USA.* Oslo: Fridtjof Nansen Institute.

Bacon, Robert, *et al.* (1990): *Demand, Prices and the Refining Industry. A Case Study of the European Oil Product Market.* Oxford: Oxford Institute for Energy Studies, Oxford University Press.

Baker and Rendall (1993): 'Pit Closures—Economics or Structure as the Cause?' in Peter Pearson (ed.), *The Economics of Pit Closures in the UK.* SEED 67. Surrey Energy Economics Centre.

Ballard, C. L., and D. Fullerton (1992): 'Distortionary Taxes and the Provision of Public Goods', *Journal of Economic Perspectives*, 6/3.

Bruce (1994): 'Heseltine's "Reprieved" Pits, Twelve Months on', *Energy Policy*, 22/5 (May 1994).

Cecchini, P. (1988): *The European Challenge. 1992: The Benefits of a Single Market*. Aldershot.

Cowhey, Peter (1985): *The Problems of Plenty: Energy Policy and International Politics*. Berkeley: University of California Press, USA.

Dahl, Agnethe (1995): National Freedom of Action in EU Environmental Policy: Denmark and the Netherlands. EED Report no. 1995/1. Oslo: Fridtjof Nansen Institute.

EBRD (1994): *Transition Report*. London: The European Bank for Reconstruction and Development.

ECON (1993): *Strukturendringer på kraftmarkedene i Tyskland og Nederland. (Structural Changes in the Electricity Markets in Germany and the Netherlands)*. Report no. 33/93. Oslo: ECON Centre for Economic Analysis.

Energidata (1993): *Evalueringen av de statlige tilskuddsordningene for ENØK. (Evaluation of Public Investment Support for Energy Conservation.)* ED report no. 93–217, Trondheim.

Energy Policy (1994): Special Issue. Markets for Energy Efficiency. *Energy Policy*, 22/10 (Oct. 1994).

Estrada, Javier, *et al.* (1988): *Natural Gas in Europe: Markets, Organisation and Politics*. London: Pinter Publishers.

Eurelectric (1991): *Quelle forme de concurrence pour le secteur electrique en Europe?* Brussels: Eurelectric.

Eurostat (1993): *Electricity Prices 1985–1992*. Luxemburg: Statistical Office of the European Communities.

French Ministry of Industry and Foreign Trade (1993): *Document d'orientation sur la politique energetique*. Paris.

Grubb, Michael (1990): *Energy Policies and the Greenhouse Effect, 1. Policy Appraisal*. The Royal Institute of International Affairs. London: Dartmouth Publishing Co.

———— (1995): *Renewable Energy Strategies for Europe*. London: The Brookings Institution.

———— and John Walker (1992): *Emerging Energy Technologies: Impacts and Policy Implications*. The Royal Institute of International Affairs. London: Dartmouth Publishing Co.

Haugland, Torleif (1993): *A Comparison of Carbon Taxes in Selected OECD Countries*. OECD Environment Monograph No. 78, Paris.

———— (1994): *The Cost-Efficiency of Financial Investment Support in Energy Conservation Policy*. EED working paper no. 1994/6. Oslo: Fridtjof Nansen Institute.

Heilemann, U., and B. Hillebrand (1992): 'The German Coal Market after 1992', *The Energy Journal*, 13/3.

Hoeller, P., and M. O. Louppe (1994). *The EC's Internal Market: Implementation and Economic Effects*. OECD Economic Studies No 23, Paris.

Horsnell, Paul, and Robert, Mabro (1993): *Oil Markets and Prices*. Oxford: Oxford University Press for Oxford Institute of Energy Studies.

Hope, Einar, Linda Rudt, and Balbir Singh (1993): *Markets for Electricity: Economic*

Reform and the Norwegian Electricity Industry. SNF Working Paper No. 12/1993. Bergen.

IEA (1985): *Electricity in IEA countries.* Paris: International Energy Agency.

—— (1987*a*): *Energy Conservation in IEA Countries.* Paris: International Energy Agency.

—— (1987*b*): *National Oil Companies and Downstream Integration.* Note by the IEA Secretariat, IEA/SOM(87)34. Paris: International Energy Agency.

—— (1988): *Coal Prospects and Policies in IEA Countries. 1987 Review.* Paris: International Energy Agency.

—— (1989*a*): *Electricity End-Use Efficiency.* Paris: International Energy Agency.

—— (1989*b*): 'World Oil Industry Structures'. Unpublished note by the IEA Secretariat, IEA/SOM(89)5. Paris: International Energy Agency.

—— (1991*a*): *Fuel Efficiency of Passenger Cars.* Paris: International Energy Agency.

—— (1991*b*): *Energy Efficiency and the Environment.* Paris: International Energy Agency.

—— (1991*c*): *Energy Policies of Poland. 1990 Survey.* Paris: International Energy Agency.

—— (1992*a*): *Energy Policies of IEA Countries. 1991 Review.* Paris: International Energy Agency.

—— (1992*b*): *Energy Policies Czech and Slovak Federal Republic. 1992 Survey.* Paris: International Energy Agency.

—— (1993): *Energy Policies of IEA Countries. 1992 Review.* Paris: International Energy Agency.

—— (1994*a*): *Coal Information 1994.* Paris: International Energy Agency.

—— (1994*b*): *Energy Policies of the Czech Republic, 1994 Survey.* Paris: International Energy Agency.

—— (1994*c*): *Energy Statistics and Balances, 1960–92.* (Diskettes) Paris: International Energy Agency.

—— (1994*d*): *Energy Prices and Taxes.* (Diskettes) Paris: International Energy Agency.

—— (1994*e*): *Energy Policies of IEA Countries: 1993 Review.* Paris: International Energy Agency.

——(1995*a*): *Energy Policies of Poland: 1994 Survey.* Paris: International Energy Agency.

—— (1995*b*): *Energy Policies of Hungary: 1994 Survey.* Paris: International Energy Agency.

—— (1995*c*): *Energy Policies of IEA Countries: 1994 Review. Paris: International Energy Agency.*

—— (1996): *Energy Balances of OECD Countries.* Paris: International Energy Agency.

International Petroleum Encyclopedia (1986 to 1994): Published annually by the Energy Group of the PennWell Publishing Co., Tulsa, Okla.

IPCC (1990): *Climate Change: The IPCC Scientific Assessment*. Cambridge: Cambridge University Press.

—— (1996): *Climate Change 1995: The Science of Climate Change*. Contribution of Working Group I to the Second Assessment Report of the Intergovernmental Panel on Climate Change. Ed. J. J. Houghton, L. G. Meiro Filho, B. A. Callander, N. Harris, A. Kattenberg, and K. Maskell. Cambridge: Cambridge University Press.

Jochem, E., and E. Gruber (1990): 'Obstacles to Rational Electricity Use and Measures to Alleviate Them', *Energy Policy*, 18/4.

Lucas, Nigel (1985): *Western European Energy Policies: A Comparative Study of the Influence of Institutional Structure on Technical Change*. Oxford: Clarendon Press.

Mandil Report (1993): *La Réforme de l'organisation électrique et gazière française*. Paris: Ministère de l'Industrie, des Postes et Télécommunications et du Commerce Exterieur, Direction Générale de l'Energie et des Matières Premières.

Mabro, R., *et al.* (1986): *The Market for North Sea Crude Oil*. Oxford: Oxford Institute for Energy Studies.

Mitchell, John (1994): *An Oil Agenda for Europe*. London: The Royal Institute of International Affairs.

Nelkin, D., and M. Pollak (1981): *The Atom Besieged: Extraparliamentary Dissent in France and Germany*. Cambridge, Mass.: MIT Press.

Newbery, David M. (1994): 'UK Energy Policy versus EU Environmental Policy', in *The International Association for Energy Economics*, Conference Proceedings, III. Oslo.

OECD (1991): *The State of the Environment*. Paris: Organisation for Economic Co-operation and Development.

—— (1993a): OECD *Environmental Data Compendium 1993*. Paris: Organisation for Economic Cooperation and Development.

—— (1993b): *Environmental Performance Reviews Germany*. Paris: Organisation for Economic Cooperation and Development.

—— (1994): *Main Economic Indicators*. (Diskettes) Paris: Organisation for Economic Cooperation and Development.

—— (1995): OECD *Environmental Data Compendium 1995*. Paris: Organisation for Economic Cooperation and Development.

Oil & Gas Journal (1985–94): Monthly, Tulsa, Okla: PennWell Publishing Co.

Peebles, Malcolm W. H. (1980): *Evolution of the Gas Industry*. London.

Percebois, Jacques (1989): *Economie de l'Energie*. Economica.

Petroleum Economist (1989–1994): Monthly, London.

Piper, Jeff (1994): 'State Aid to the European Union Coal Industry', in *Energy in Europe*, 23 (July 1994). European Commission.

Radetzki, Marian (1994a): *Poland's Hard Coal Industry: Prospects after Completed Restructuring*. SNS Energy, Occasional Paper No. 58 (June 1994).

—— (1994b): 'Hard Coal in Europe: Perspectives on a Global Market Distortion'. *OPEC Review*, 18/2 (Summer 1994).

Richter, Joerg-Uwe (1992): 'Energy Issues in Central and Eastern Europe: Considerations for International Financial Institution', *The Energy Journal*, 13/3.

Ross, Petter R. J. (1987): 'The Government as Entrepreneur: With Special Reference to the United Kingdom'. Chapter 13 in Khan, Kameel, *et al.* (1987), *Petroleum Resources and Development*. London and New York: Behaven Press.

Shaffer, Ed (1983): *The United States and the Control of World Oil*. Croom Helm.

Sandnes, Hilde, and Helge Styve (1992): *Calculated Budgets for Airborne Acidifying Components in Europe 1985, 1987, 1988, 1989, 1990 and 1991*. Oslo: The Meteorological Institute, under the EMEP co-operative programme.

Schiffer, Hans-Wilhelm (1988): *Energiemarkt Bundesrepublik Deutschland*. Cologne: Verlag TUV Rheinland GmbH.

────── (1994): *Energiemarkt Bundesrepublik Deutschland*. Cologne: Verlag TUV Rheinland GmbH.

Schipper, L. (1987): 'Energy Conservation Policy in the OECD: Did they Make a Difference?', *Energy Policy*, 15/6 (December 1987).

────── and S. Meyers (1992): *Energy Efficiency and Human Activity: Past Trends and Future Prospects*. Cambridge: Cambridge University Press.

────── M. J. Fugueroa, L. Price, and M. Espey (1993): 'Mind the Gap: The Vicious Circle of Measuring Automobile Fuel Use', *Energy Policy*, 21/12.

Schwarz, Michel, and Michael Thompson (1990): *Divided we Stand: Redefining Politics, Technical and Social Choice*. New York: Harvester Wheatsheaf.

SFT (1994): *Forurensning i Norge 1994 (Pollution in Norway 1994)*. Oslo: Norwegian Pollution Control Authority.

Stern, Jonathan P. (1984): *International Gas Trade in Europe: The Policies of Exporting and Importing Countries*. The Royal Institute of International Affairs. London: The Brookings Institution.

────── (1990): *European Gas Markets: Challenge and Opportunity in the 1990s*. Dartmouth: The Royal Institute of International Affairs.

────── (1995): *The Russian Natural Gas Bubble: Consequences for European Gas Markets*. Washington, DC: The Royal Institute of International Affairs and The Brookings Institution.

van Aalst, R. M. (1993): 'Atmospheric Pollution: A European Concern', in Group of Sesimbra (ed.), *The European Common Garden*. Brussels. GLOBE (Global Legislators Organization for a Balanced Environment).

World Energy Council (1994): *New Renewable Energy Resources: A Guide to the Future*. London: World Energy Council.

Yergin, Daniel (1991): *The Prize*. New York: Simon & Schuster.

INDEX